Cosmic Ships

COSMIC SHIPS

TRUTH AND LIES
ABOUT UFOS, OTHER
HUMANITIES, AND
OUR FUTURE

SAMAEL AUN WEOR

GLORIAN

2010

Cosmic Ships
A Glorian Book / 2010

Collected from writings by Samael Aun Weor, originally
published in Spanish.

This Edition © 2010 Glorian Publishing

Print ISBN: 978-1-934206-39-3
Electronic ISBN: 978-1-934206-58-4

Glorian Publishing is a non-profit organization. All
proceeds go to further the distribution of these books.
For more information, visit glorian.info.

Contents

Editor's Introduction ix

Chapter 1: Concerns for the World 1

Chapter 2: The Naked Truth 51

Chapter 3: New York's Blackout 57

Chapter 4: Flying Saucers and
  Little Green People 71

Chapter 5: Pure Science 77

Chapter 6: Flying Spheres 81

Chapter 7: Cosmic Ships 89

Chapter 8: A Mexican Traveled to Venus 99

Chapter 9: The Pluralized "I" 113

Chapter 10: A Jupiterian Visitor 121

Chapter 11: The Gnostic Movement (1964) ... 129

Chapter 12: The Extraterrestrials' Mission 137

Chapter 13: Answers about Extraterrestrials .. 159

Chapter 14: The Consequences of the
  Comet Kondoor 167

Open Letter 187

Glossary 197

Index 221

What science rejects today, it accepts tomorrow.

- Samael Aun Weor, *The Perfect Matrimony: The Door to Enter into Intitiation* (1950)

Editor's Introduction

For more than fifty years, evidence has been accumulating that there is more to life in our corner of the universe than just fast food, taxes, and alcohol. From across the spectrum of society have come reports of ships in the sky, unexplained lights and movements, and strange occurrences. Military officers, government officials, police officers, astronauts, and common people from all walks of life and countries around the world have reported seeing ships that demonstrate abilities far beyond our level of technology. Yet, governments remain silent or deny any knowledge, while the media crucifies those who come forward with their experiences.

Skeptics claim that sightings of unidentified flying objects (UFOs) or extraterrestrials are evidence of mass hysteria and collective delusion, of people seeing what they want to see. In many cases, this is undoubtedly true. Yet, such causes do not explain all cases and sightings. There are too many cases and too much evidence. This book is not a catalog of such cases, or an attempt to explain them. Moreover, this book is not an attempt to convince the skeptics. This book collects writings by Samael Aun Weor published in Spanish between 1950 and 1977. In 1950, Samael Aun Weor published his monumental first book, *The Perfect Matrimony: The Door to Enter into Initiation.* It included a chapter† about "flying saucers," making him one of the first authors in the world (and likely the first in Spanish) to write about the appearance of strange craft in the skies of the world.

In the decades since UFO sightings began to be widespread, the public has been subjected to an intense campaign from governments and media organizations, intended to discredit and ridicule any and all information about extraterrestrials and unidentified flying objects. Their campaign has been a resounding success, and made more effective by the rapid spread of ridiculous "new age" theories and beliefs, and the abundance of very unbalanced people making a huge variety of outrageous claims about extraterrestri-

† - In later editions of *The Perfect Matrimony,* he extracted that chapter to form the basis of the book you have in your hands; that chapter is here as Chapter Six.

al life. The result of all of this is that now, anyone who affirms the existence of life on other planets is immediately suspected of being a "new age" buffoon, a crackpot. People everywhere chuckle at the idea of superior beings visiting this planet. Now, people hesitate before mentioning the subject. This is how effective the propaganda has been.

Nonetheless, illusions cannot withstand the presence of the facts, whether those illusions are created by narcissistic spiritualists, government agencies, or Hollywood. The facts demonstrate that something is happening on this planet that our governments and media organizations are not willing to discuss. The facts demonstrate that we are not the only conscious beings in the universe or on this planet. Such facts can be found in the hundreds of books and videos documenting sightings by reputable people of extraterrestrial ships and related events that have been happening for decades. Yet, such facts can also be acquired through our own experience, and not just by accident or fate. Truly, we cannot know the truth of anything unless we experience it for ourselves.

Gnosis, the term used by Samael Aun Weor to describe the root wisdom at the base of all existence, is a Greek word that means "knowledge acquired through experience." Such knowledge is based on the facts of experience, not mere theory, supposition, or belief. His writings are based on facts he confirmed through his own experience. It is important to keep this distinction in mind, because nowadays, most of what we read was dreamed up by the writer or merely repeats what other writers said. The writings of Samael Aun Weor are different. He said,

> ...there are authors who write marvels, but when one looks at them, one realizes that they have not lived what they have written; they did not experience it in themselves, and that is why they are mistaken. I understand that one must write what one has directly experienced by oneself. For my part, I have proceeded in this way. - *Tarot and Kabbalah* (1977)

Samael Aun Weor wrote over seventy books and gave thousands of lectures

about the practical, experiential way
to fully develop all the potential of the
human being. That is, he worked on
behalf of humanity, in order to freely give
everyone a solid introduction to the real
method to transcend suffering and realize
the purpose of our existence.

The basis of the Gnostic method is a
very rigorous self-examination to dis-
criminate between the illusions we have
about ourselves and the actual facts. The
undeniable proliferation of crackpots and
boastful self-proclaimed "experts" (whether in science or spirituality) is caused by
self-delusion. People are hypnotized by
their self-projected illusions. People are
willingly blind to the facts of their lives,
and instead only want to see what gives
them a sense of psychological security.
The scientist only sees what supports his
theories and his identity as a "scientist."
The religious one only sees what supports
his beliefs. The worker only sees what
makes his efforts seem virtuous and good.
None of us are willing to see anything
that contradicts our sense of self. We do
not see the facts. Our perceptions are
filtered.

The basis of Gnosis is to learn to see ourselves and our environment without any filters. This is the very definition of real science: to see what is actually there, without judgment, theory, definition, or explanation. The true Gnostic sees things as they are.

Naturally, everyone believes that they see things as they are. No one is willing to admit that they are completely asleep, unaware of reality. Yet, this is the truth. Only the one who is becoming aware that he is asleep is on the way to awakening.

In order to understand this book, you will need to adopt a new way of seeing. Instead of reading this book through the filters of what you have been told or believe, read it with an open mind, receptive to the idea that perhaps what you have been told is wrong. Perhaps we have been wrong about more than we imagine. Certainly, judging by the state of our world, we do not have much evidence to claim that we are all that bright. It is beneficial to open our hearts and minds to learn how we are mistaken, for the one who refuses to see his mistakes is condemned to repeat them.

Chapter 1

Concerns for the World

FROM A LECTURE

Ladies and gentlemen, tonight we are going to speak about some very interesting subjects of living reality. Obviously, we are in times of extraordinary concerns for the world. Thus, it is necessary—if it is what we want—to deeply reflect on them at this present moment in which we live.

It appears that we have a powerful modern civilization, since many advances have been made in the fields of physics, chemistry, medicine, engineering, etc. In this, our great civilization, we have built powerful ships and directed them towards the Moon. Ships have scanned space, and have landed on the lunar surface, etc. Ships have also been sent to Venus, although they were not manned. Excursions to Mars are planned, and it has been said that around the year 1985, or maybe later, the United States is planning to send a manned nuclear rocket to Mars. We will wait for concrete results on this particular subject-matter. All of this is intriguing and extraordinary in its depth.

Television has fulfilled a great mission, since thanks to the television we perfectly followed the progress of the rockets that descended upon the Moon. Much about lunar life was learned, and this was abundantly investigated.

Many wise men previously thought that the Moon was a piece of Earth projected into outer space, yet the carbon-14 tests were definitive: Earthlings arrived at the logical conclusion that the Moon existed before the Earth. The Moon is older than the Earth, and this by itself is sensational. All those wise men that in the past maintained the theory that the Moon was a piece of the Earth projected into outer space were lamentably mistaken. I repeat: tests and rigorous analysis of the lunar rocks by means of carbon-14 indicated to Earthlings that those wise men who in the past maintained the thesis that the Moon was a piece of the Earth were mistaken.

Therefore, we are in times of great scientific concerns for the world. We must reflect deeply upon all of these things, at least for awhile.

You are here because you have scientific concerns for the world, and I am here because I also have great scientific concerns for the world. You have come willingly to listen to me, and I am prepared to converse with you; thus, between us there must be an exchange of ideas. Indeed, we have all met together in order to study many things, in order to analyze diverse important subjects that are interesting to you and me. Thus, I want all of us, all together, to analyze these subjects of living reality.

Obviously, we Earthlings struggle for the conquest of space, and we are sincerely doing it. Our scientists fly upon the wings of their projects towards a future in which the Earthlings definitively conquer other worlds. Nevertheless, we must not in any way become fascinated with so many sophisms. Instead, it is convenient that we—by our own means—investigate for ourselves. In this way, we will probably avoid many disappointments.

You know that in these times, much has been spoken about the subject of cosmic ships coming from other planets. There is a type of antinomy, a very interesting antithesis between rockets launched by "Tyrians and Trojans"† to the Moon or Venus, and cosmic ships originated from other planets. Among the "Tyrians and Trojans" there is a certain skepticism that leads to nowhere. There are therefore concerns on one side, and on the other—antagonistic concepts, clashing opinions; to reflect upon all of this is praiseworthy.

When we hear about flying saucers, we may pay attention or we just smile a little, skeptically, but listen: there is something real in all of this, since it does not seem to me—in any way—that our planet Earth is the only inhabited planet.

When one studies the "Panspermia" of Arrhenius, one discovers with mystical astonishment that the seed-germs of life come from other planets; thus, the theories of Arrhenius are intriguing.

Obviously, within the luminiferous dust of stars, our planet Earth is an inhabited world, a world that rotates around the Sun, a planet like any other in the infinite space. The law of philosophi-

† Editor's Note: The Tyrians and Trojans are the two cultures in conflict with one another in Virgil's *Aeneid*.

cal analogies invites us to think that if our planet Earth is inhabited, then there also must be other inhabited planets within the infinite space. I will never think that the seed-germs of universal life are an exclusive patrimony of the planet Earth; it seems to me that exclusivism in this sense is regressive, reactionary, retarded.

I invite you to think: if we are fighting for the conquest of space, it is then possible that this same fight also exists on other planets. Therefore, I will never discard the idea of the possibility of extra-terrestrial people, inhabitants of other spheres, who have already conquered outer space. To think that we are the only ones in such a huge space, made up of millions and millions of planets, is too reactionary and snobbish.

Remember that in the times of Columbus, many were those who laughed at that wise man, that great navigator, when he hurled, as they said then, through the ocean, beyond "Cape Finisterre." Then, in the time of Columbus, there was the belief that the Earth was flat, square, thus nobody in Europe accepted the possibility of life beyond Cape Finisterre, which in Latin means "where the Earth (terre) ends (finis)."

It appears that sometimes there are those who think with a medieval mind, denying the possibility of conscious and intelligent life on other planets. Indubitably, they think with an old fashioned, anti-revolutionary, medieval criterion.

Let us admit the possibility of life on other planets.

The cosmic ships are a reality. People more cultured than us exist on other inhabited planets. They already conquered outer space, and about this I can give you a convincing testimony.

If I were to base myself on mere intellectual lucubrations, then indeed I would not have a basis to affirm the thesis of other planets inhabited by extraterrestrial people. If I were to base myself solely on purely intellectual conceptions of formal logic or dialectic reasoning in order to emphasize the idea of the possibility of the existence of extraterrestrial people, then I would be just another theoretician. Yet, truly, the existence of extraterrestrial beings is a fact for me. I know them personally, in flesh and bone, and for that reason I do not have any inconvenience in giving testimony about them. If you believe it, good. If you accept it, wonderful. If you reject it, that is your business. In any case, I will give my testimony.

One day, it does not matter which, when residing in Mexico City, I had to visit "El Desierto de los Leones" (The Desert of the Lions National Park). I wanted to peacefully abide there, even if only for a few hours. I wanted to deliver myself to the calmest of reflections.

The Desert of the Lions National Park is a protected wilderness area measuring seven by three kilometers.

All of a sudden, I felt myself attracted towards a certain place in the forest. I saw a space there, in the woods. I do not know why I had the feeling of directing myself towards that place, even when indeed I found an enormous cosmic ship standing upon a steel tripod.

Obviously, I confess to you that I felt completely confused, moved; that discovery left me absolutely astounded. However the story does not end here: a metallic hatch opened, and I saw the chief or captain descend from that ship... the crew came behind him. Naturally, I addressed the chief, the captain; I saluted him, and he answered my greeting in perfect Spanish...

"Good morning," I said to him.

"Good morning," answered the captain...

Amongst the crew I saw two elderly ladies. What age could they be? I do not know. Unquestionably, their ages would have to correspond to other times, not to our earthly time.

I spoke to the captain, saying, "Sir, I would like to know the planet Mars, since my own spiritual, divine, sparkling seed-germ relates to that planet of the infinite space." (My Monad, we would say, speaking in the style of Leibnitz, who occupied himself so much with the Monads)...

After some minutes, the captain in charge of that ship took the floor and said, "To Mars is what you said?"

"Yes, I would like to know the planet Mars, and I would like you to take me there. I am willing to go with you now, immediately; nothing can hold me on the planet Earth."

"To Mars," said the captain to me, "such a planet is just there; indeed, that world is very nearby."

Thus, when speaking in this manner, I comprehended that my request, or that my pretension had been too poor. I believed I had requested something very great, but so... why lie? My request had been indeed very poor... By certain intuitive signs, they made me understand that the ship that seemed so splendid to me came down from a mother ship that was hidden, orbiting the Earth.

Our solar system—well-known with the name "solar system of Ors"—was not in any way unknown to the captain, yet it was but one of so many solar systems of the unalterable infinite... Undoubtedly, I found myself before intergalactic travelers, people who travel from galaxy to galaxy, wise and cognizant individuals.

"I am a writer," I said to him, "and I would like to be taken to other inhabited planets, in order to write and give convincing testimony to this humanity about the existence of other inhabited planets... I am a man," I said to him, "I am not a simple intellectual animal; the request that I make of you is not for me, but for this humanity in general. I would like in some way to contribute to the general culture of this world on which I live..."

In short, I set out many concepts; nevertheless, that captain kept silent. I even held on to that tripod of steel, with the purpose of not letting go of it until they agreed to put me inside the ship and take off with me. But everything was useless: they kept silent...

I examined that man and all the crew: they were personages of a copper color, ample forehead, thin body, stature of about one meter with twenty, thirty, or forty centimeters [3.8 to 4.4 feet], at the most...

The crew finally sat down on some wood trunks that were in the forest. The ladies were two venerable elders, and I could not do less than to observe such strange creatures...

I could not see in them our terrestrial perversity. I carefully noticed the sense of human responsibility that they had. They spoke little, because they have a very high concept of the word. They do not speak on a whim like us: they speak little and say much. For them, the word is gold, powdered gold. They only use it in very indispensable cases...

I did not see the face of assassins on them, like ours, the Earthlings, neither did I see that Machiavellian look on them (with which so many particular films are adorned). In those strange creatures only shone wisdom, love, and power. They are humans, but true humans, in the most complete sense of the word.

None of them wanted to abduct me. On the contrary, I fought too much, requesting from them that they take me. I am sure that if they had granted such a request to me, in no way would they have made me a "guinea pig for their laboratory." We Earthlings are another thing. If by any chance we managed to catch an extraterrestrial, it is certain that he would immediately go to the laboratory, and as far as the ship, we would confiscate it, and with it, like a pattern, we would be able to build many other ships in order to bomb defenseless cities, in order to conquer other worlds by force, and make devilish things and more, because we Earthlings—beginning with me—are truly terribly perverse; that is the crude reality of the facts. In no way I have come here in order "to wash my hands" in front of you, and to say that I am a meek sheep; no! All of us here are "cut with the same scissors," thus the defects that I have, you have them also, and vice versa...

Therefore, I assure you that the testimony that I give about those people is sincere, truly sincere. I am not trying in any way to deform the testimony, to deform the truth.

Finally, from those of the crew who were seated upon the wood trunks there, one of the ladies stood up, and in the name of all the crew she took the floor and said, "If we place a plant that is not aromatic next to another that is aromatic, the one that is not aromatic would be impregnated with the aroma of the one that is aromatic."

Soon she continued, "The same happens on inhabited planets. Worlds that previously advanced poorly, with perverse humanities, were little by little transformed by the aroma, the vibration, of neighboring planets. But, as you see, we just arrived here at this planet Earth, and we do not see that the same happens here. What is happening on this planet?"

Well, the question that they asked me was tremendous, and I had to give an answer, then, of high quality... Thus, without reflecting that much, but of course taking care of the word very well, I said, "This planet Earth is a mistake of the Gods..." But soon I completed it, clarifying the concept as best I could, and said, "This is how the Karma of the worlds is."

Karma is a word that represents or means cause and effect: by such a cause, such an effect. The Earth has causes that brought it into existence, and if within those causes are mistakes, more or less, the effects will have the outcome of those mistakes" ...

Thus, when saying, "This is how the Karma of the worlds is," with great astonishment I saw that the elderly lady who had spoken agreed, inclining her head with one respectful bow. She did not say anything, but simply agreed. The other elderly lady did the same: she made one respectful bow, and all of the crew, in moderate genuflection, agreed.

Well, I will say something to you: I thought that they were going to pull me by my ears, because it was terrible for a poor devil like me to give an answer to people who travel from galaxy to galaxy, but I did it. I did it, my answer worked, and that cheered me...

Of course, I resolved to take the best advantage of such assent. Thus, I said to myself, "Well, this is the moment," thus, I returned to reiterate my request to be taken to another planet of the infinite space, in order to give testimony to the people about the reality of other inhabited planets...

"I am writer," I said to them. "And it is not for me, it is for this humanity; take me..." To no avail. My requests were worthless.

The silence was terrible. Finally, the captain pronounced a phrase, nothing but that, because they speak little and they say much. They never utter the word if they are not going to fulfill it. They are not like us; i.e. we tend to say to a friend, "Tomorrow we will meet in the morning at nine in the café so that we can converse about the business," and the friend does not arrive, and if he arrives, he appears around ten, or eleven, or twelve...

So, those people speak little and say much. It seems as if those personages were truly Gods with human bodies (they gave me that impression when conversing with them)...

Thus, I obtained an answer, and as soon as they gave it to me, it is clear that I was satisfied, "Along the way," said the captain, "we will see..." Nothing more, that was the only thing that he said to me, but that for me was definitive. If an Earthling had said the same to me, I would simply have considered those words like an escape, like an evasion, for example as when one asks a job, and they say to you, "We will consider you when there is a vacancy." When receiving such an Earthling-answer, one feels like leaving, running five hundred kilometers per hour, since we can be sure that we have failed in the request...

Yet, I was not conversing with Earthlings, but extraterrestrials. "Along the way, we will see." Which "way" was that captain talking about? The esoteric path of initiation, the path that I am following, and that many are following; the path that leads to the Superman: that straight, narrow, and difficult path of which the Christ speaks, that mysterious path tread by Dante, Hermes Trismegistus, or Jesus of Nazareth. I follow such a path; therefore, the words of that captain filled me with strength.

Well, he gave me his hand (his right hand), then he boarded the ship by a stairway. Also, those of the crew boarded. I comprehended that I should withdraw, thus, I did so. I did not want in any way that my body be instantaneously disintegrated by the force of that ship. Thus,

I withdrew a certain distance so that I could observe, through the trees, the moment in which the ship took off. It raised slowly, until a certain point, and soon it hurried through the infinite space, without making any type of noise...

I assure you that I am giving a testimony about people who already conquered space, about the extraterrestrials.

I have come here to tell you the truth and nothing but the truth. I have not come to give false testimonies, because with that I would not gain anything, nor would you gain anything; I would deceive myself and I would commit the absurd crime of deceiving my fellowmen. I am giving you a testimony of the truth, of what concerns me about the extraterrestrials. If you believe me, wonderful; if you do not believe me, it does not matter to me. If you laugh, well, that is your business, yet in any case, Victor Hugo in one of his works stated, "The one who laughs at what he does not know is an ignoramus who walks on the path of idiocy." So, I give my testimony; the rest is up to you.

Yes, there exist other people who already conquered outer space, and who are not Earthlings. They are people who come from other densely populated planets. It is urgent to comprehend that those people—who already conquered the infinite space—do not have vices. They do not drink, they do not smoke, they do not fornicate, they do not adulterate, they do not rob, they do not kill. They are perfect, in the most complete sense of the word... Thus, I say to myself and to you, thinking aloud now: Have we, Earthlings, such merits? Are we worthy to conquer infinite space? And if we could attain it, what would be our conduct on other inhabited worlds? Are we sure that we are not going to go and drink there, to get drunk, to adulterate, etc.? Are we so perfect that we believe we are capable of conquering infinite space?

Now then, I understand that those cosmic ships are multidimensional. It seems to me that the three dimensions—length, width, and height—are not everything. Euclid's three-dimensional geometry has been abundantly discussed. I.e. this table has length, width, and height, it has three dimensions, but there must be a fourth vertical in this table. What would it be? I say that this is time: how long has it been since this table was fabricated? Behold the fourth vertical there.

Indubitably, the fifth coordinate also exists; I understand that it is eternity.

And beyond the fifth dimension, there must exist the sixth, that is to say, a dimension that is not time, neither eternity, nor the three-dimensional world. The fifth coordinate is eternity, the fourth is time, but what would the sixth be, and the seventh? The sixth is beyond eternity and time, yet, as far as the seventh, it is the dimension zero, unknowable (pure Spirit, we would say).

Indubitably, there must be seven basic, fundamental dimensions. However, while we base our existence on the three-dimensional dogma of Euclid, we will remain in a regressive, retarded state... At the present time, modern physics is retarded, regressive, antiquated, old fashioned, because it is exclusively based on the three fundamental basic dimensions, of Euclid's three-dimensional dogma.

Extraterrestrial ships are based on a different geometry. I say that a tetra-dimensional geometry must be created; this would be possible if we more thoroughly investigate the atom. Obviously, it is in the atom where the fourth vertical is drawn up.

The day in which we can draw up the fourth vertical on paper, we will be capable then to also create a tetra-dimensional geometry. With a geometry like that, we could then build ships of four dimensions, ships able to travel in time, now to a remote past, now to a remote future. With ships like that, we could conquer the infinite space. Regrettably, we cannot create those types of ships yet.

In order to travel to Mars in a nuclear rocket, we will take about two years, and according to the explanations of those extraterrestrials that I met in "El Desierto de los Leones" I understood that in less, in a matter of minutes, they are on Mars—for them Mars is "just there," at the corner store, so to speak—and it is because they put their ships within the fourth vertical. Their ships are propelled by solar energy, and this is wonderful.

Nevertheless, we Earthlings needed to send rockets equipped with liquid fuel, and our astronauts performed fifty thousand acrobatics in order to land on the

Moon. However, extraterrestrials do not need such acrobatics, since for them the Moon is "just there." Therefore, I do not see why we have to feel so proud of our so-boasted modern civilization.

I invite you to comprehend that we Earthlings are nothing but embryos, and that our so-boasted modern civilization is not really that worthy. I invite you to thoroughly comprehend this subject of the conquest of interplanetary space.

It is necessary to analyze, it is necessary to study. It is necessary to comprehend that if we want the conquest of infinite space, we must begin to study ourselves, because the laws of the cosmos are within us, ourselves, here and now. If we do not discover the laws of the cosmos within us, we will never discover them outside of us.

The human being is contained within Nature, and Nature is contained within the human being. Therefore, if we want to conquer infinite space, we must begin to conquer ourselves.

At the present time, we are victims of circumstances. We have not learned how to handle the diverse circumstances of life. We still do not know how to determine circumstances. We are toys of all the forces of the universe.

We live in a convulsing world, a world that is going to pass through great catastrophes. Earthquakes are coming. They have been walking across America from south to north. One day, Chile is affected by great earthquakes and tidal waves; later Caracas, followed by Colombia. Nicaragua was shaken, followed by Honduras, and in Guatemala, earthquakes just happened. It is necessary to know that soon all of our cities in Mexico will be shaken by earthquakes.

San Francisco, California is called to disappear. There is a fault at the foot of the California peninsula that has already been studied. It is a deep crack that has already begun to devour California little by little. Obviously, California will sink to the bottom of the Pacific Ocean.

We live, then, in a world that is threatened by great convulsions, and the psychological state in which we are, and that of our civilization, etc., deserves to be reflected upon by us a little.

The bottoms of the Atlantic and Pacific Oceans are full of deep cracks. In the

Pacific, mainly, there are some cracks that are so deep that they already put the fire in contact with the water. The water of the ocean penetrates within the interior of the Earth in those zones where the liquid fire exists, thus this is forming pressure and steam that increase from moment to moment. The pressure and steam are originating earthquakes on a great scale, and all of you, distinguished gentlemen and ladies, are going to be convinced within a short time that there will be not a single place on the planet Earth where one can be safe.

The earthquakes and tidal waves must intensify, due to pressures and underground steam.

The ice of the North Pole is melting, and because icebergs float, they drift with water currents towards the equator into warmer water.

Hot currents of water are being produced at the South Pole. These currents are exiting through some craters. These hot water currents penetrate certain places of Guinea.

There are changes within the planet Earth, and if the pressures and steam continue, one day the terrestrial crust will erupt. There is no doubt that at the present time, any cosmic event, i.e. the arrival of some gigantic world, would be enough to produce such an explosion.

We are seated over a powder barrel, and we do not realize it. The Earth in its entirety is being prepared for formidable geologic changes. Thus, Nature at the moment is passing through difficult processes. Nature is experiencing a great agony. The fire of the interior of the Earth is uneasy.

Sadly, we, on the epidermis of this planet, believe we are very safe. We are raising powerful buildings, as if they will never fall to the ground. We build powerful ships, as if they will allow us to flee to other planets at any given moment. Yes, we feel like masters of the universe, yet regrettably, any stomachache is enough to put us in bed. Yes, we are weak, yet we believe ourselves to be invincible.

It seems to me that we must reflect upon what we are, upon what is happening, upon what is happening at this moment...

Two frightful world wars have occurred in this twentieth century: the world war from 1914 to 1918, and the world war from 1939 to 1945. And, there will be a third world war, and it will be atomic. Then, there will be a great holocaust: powerful cities will be reduced to ashes. Millions of people will perish. The gravest of all of this is that the abuse of atomic physics will take us to disaster.

A day will arrive on which the decomposition of the atom chain will occur, and then the scientists will not be able to control the atomic energy. There is no doubt that the radioactive contamination will be frightful; i.e. clouds loaded with radioactivity will pour over the harvests, contaminating them. Therefore, during the third world war we will no longer have the necessary food to eat, because the radioactivity will have contaminated the harvests completely. The contaminated food will be worthless for our nourishment.

At the rate that we are going, we should not feel very safe with a civilization that staggers, and we should not feel very safe with our theories, concepts, or ideas, either. It is worth the trouble for us to review everything we have learned in school, college, university, in books written by different writers.

I am not trying to attack any theory, no. I am only inviting all of you to reflection, and nothing else; this is the only goal of this lecture.

There is a law known as the law of universal entropy. If we take two full water kettles, one containing hot water and the other containing cold water, and we place them together, we will see in them a devolving disorder. Behold here, universal entropy.

If people do not work upon themselves, if they do not try to pass through a type of psychological revolution, if they do not modify their customs, their way of life and way of being, then they will march in accordance with the law of entropy. They will devolve over time, and a day will arrive in which there will be no difference between person and person. All of us will become terribly perverse.

As far as the planet Earth is concerned, we cannot deny that it is subject to the law of entropy. The atmosphere is completely contaminated, the seas have become enormous garbage containers, and many marine species are disappearing. Fish have died in the rivers. To find a river that is not contaminated is already difficult. The fruits of the Earth have been adulterated with grafting upon grafting. Now, it is difficult to eat a legitimate apple. Now we must eat "grafted apples"... All of this has altered the order of the universe, the order of Nature. Thus, there are soils that no longer produce.

At the moment, the globe has four thousand five hundred million people, and the food supply will not be of sufficient quantity to maintain so many people. In the coming years there will be millions of people who will die of hunger. Even at this moment, there are plenty of people who are perishing because of hunger.

Therefore, the Earth as a whole is perishing. It is progressing according to the law of universal entropy. The fields that once were cultivable, that bore fruits in abundance in order to maintain everybody, are now sterile. The experiments made with atomic energy and those chemical fertilizers have sterilized the fields. Everything moves in a devolving way.

At these moments, the Earth is in agony, and what is worse is that it is in agony and we do not realize that it is in agony. Obviously, if a person is in agony, we already know what will happen to him. Similarly, if our planet Earth is in agony, we must understand what will happen to it.

A day will come in which the Earth will become equalized everywhere, turned into a gigantic Sahara, or better said, turned into another moon of the infinite space.

Nevertheless, the wisdom of the creative Demiurge of the universe is magnificent. It is not irrelevant to emphatically tell you that transformation is only possible by means of sacrifice. I.e. if we did not sacrifice the coal in the steam engine, we would not have the steam-power to move the train. Similarly, we will say that by means of a great sacrifice, the transformation of the world will also be possible.

We know well that the axes of the Earth are rising up; the day on which the poles will become the equator is not distant. The day in which the equator will become poles is not distant, either. When this occurs, the seas will change their beds and will swallow the whole planet; there is no doubt that a great chaos will come...

Again, at this moment the ice of the North Pole is already melting. This is originating enormous hurricanes that are devastating entire cities and causing damage, as the hurricane that recently caused many terrible things and obliterated Honduras.

So, the icebergs are now moving towards the equatorial zone. No longer is the magnetic pole coinciding with the geographic pole. If at these moments an airplane takes off directly towards the North Pole, guided by a compass, and if it landed exactly on the magnetic pole, the pilots would find with astonishment that the geologic pole is no longer there: the geologic pole is rotating away; it is going towards the equator.

So the magnetic pole and the geologic pole no longer coincide. This causes the climates to change, starting with certain disorders in the seasons, mainly in the spring and summer. This causes the seas to rise from their bottoms, and this powerful civilization that we have created will be destroyed. What is gravest of all of this is that with it we will also be destroyed. We will also perish.

The ancestors of Anahuac said, "The Children of the Fifth Sun"—talking about us—"will perish by fire and earthquakes..." Now this is properly determined with the catastrophe in Guatemala (that, between parentheses, was very serious, since it not only trembled, but continues trembling in that unfortunate country, and the dead are increasing).

Therefore, this humanity will perish by fire and earthquakes, and finally will be definitively wiped from the face of the Earth, when the oceans leave their beds. Thus, after this tremendous and frightful sacrifice, someday from the midst of chaos, new continents will arise, where a new humanity will live. Virgil, the great poet of Mantua, said, "The Golden Age has arrived and a new progeny commands..."

Yes, we are so perverse that we brought about atomic wars, but the day will arrive in which a peaceful humanity will live on the face of the Earth, a humanity filled with love, an innocent and pure humanity, a beautiful and wise humanity.

So then, the planet Earth once emerged from the consciousness of that which is called God, from the ineffable divine, which is where we must return now. But until now, we have marched on the path of perversity, thus we will have to perish. But as Peter said in his Epistle to the Romans:

> *Looking for and hasting unto the coming of the day of God, wherein the heavens being on fire shall be dissolved, and the elements shall melt with fervent heat?*
>
> *Nevertheless we, according to his promise, look for new heavens and a new earth, wherein dwelleth righteousness.* - 2 Peter 3: 12, 13

There will be new heavens and a new Earth and in them also a new humanity will live.

Making reconsiderations on all of these principles, it is reasonable and worthwhile that we fight for a radical transformation. It is worth suffering the trouble to make a new man within us.

We do not know ourselves, and we need to know ourselves, since within us there are wonders that we do not know.

Somebody said to me the other day, "Sir, indeed I know myself."

"It pleases me," I responded to him, "that you know yourself, but answer the following question: how many atoms are in a single hair of your mustache?"

When asking him this question, he kept silent, and finally he exclaimed, "That, I do not know."

I said to him, "If you do not even know how many atoms are in a single hair of your mustache, how do you dare to affirm with great emphasis that you know yourself in a totally integral manner?" The man remained confused.

Within us, there is something more than the physical body: there is a psychology that we must study. The physical body is not everything. You feel attracted towards your physicality. You know that you have a body of flesh and bones because you can touch it, because you can feel it, but scarcely can you admit that you have a psychology, because indeed it cannot be felt physically.

When somebody admits that he has his own psychological, particular, individual idiosyncrasy, in fact he begins to self-observe. Obviously, when somebody self-observes, he begins to become different from others, and has possibilities for changing.

A nucleus of people has to be saved from this humanity, people who will change, people who—ahead of time—will attain a psychological change. Such people will be helped and taken to a certain place in the Pacific Ocean, and thence they will contemplate the duel of the water and the fire over centuries. And finally, when new lands arise from the bottom of the oceans, those people that have changed will be able to live peacefully; they will become the nucleus of a future humanity.

We need to change, and we cannot change if we do not psychologically self-observe. For that reason I said that when somebody begins to self-observe psychologically, he provides hope for change, and to become a different person.

We need to self-observe our thinking, feeling, and acting.

It seems to me that psychological observation is not a crime. It seems to me that to attempt psychological change is not a crime.

The factors of discord that produce wars in the world exist within ourselves, within our persona. In these times, much is spoken about peace, i.e. Mussolini said, "Peace is an olive branch on the sharp edge of eleven million bayonets." Behold his kind of words and concepts. The Italians executed him; they applied to him their famous "Italian vendetta," giving him punches and kicks. Finally his corpse fell to the ground. A quite sadistic citizen, observing the corpse of "Il Duce" in the mud, exclaimed, "Il Duce has become a pig..."

Peace is not a matter of propaganda, pacifications, the UN, nor pro-peace armies, etc. Remember that the UN has sent armies to fight for peace. Do you believe that fighting for peace is peace? You yourselves are witnesses of the UN armies that have attacked other armies. The UN has bombed, has taken up the rifle. Do you believe that this is how one works for peace?

While the factors that produce wars continue to exist within us, there will always be wars in the world.

Fear is one of the main causes for world-wide armaments. If a man fears another man, he arms himself with a weapon, a pistol at his waist. Why? Because he fears the other. If he did not fear him, he would not carry a pistol.

If a nation loads itself to the teeth with weapons, if it acquires atomic bombs, ultramodern cannons, etc., it is because it fears that another nation will invade, it fears that another nation will attack.

Fear is the cause for many injustices to be committed. A man kills another because of fear. Fear for one's life causes many to become thieves. Fear of hunger causes many women to prostitute them-selves. Therefore, while the factors of fear, of fright, continue to exist within us, there have to be wars, prostitution, rob-bery, murders, etc.

If we want to fight for peace, we must end the factors that produce wars. Fear is one of them. Do we want peace? Well then, we must finish with egotism, since each one of us says, "First I, second me, and third, myself." If such egotism is projected worldwide, if the nations say, "First I, second me, and third myself," then there always will be conflicts of interests between country and country, and war will be unleashed.

Therefore, peace is not a matter of pacification, propaganda, nor of armies of peace, nor of the UN, nor of UNESCO, nor of OAS, since if the factors that produce wars continue to exist within us, there will always be wars in the world.

Peace is an ineffable atomic substance that is beyond good and evil, and that emerges from the abstract absolute space.

It is necessary that we explore ourselves in these moments of worldwide crises and bankruptcy of all principles. It is necessary that we observe ourselves psychologically.

At these moments in which the Earth is convulsed by earthquake upon earthquake, it is necessary that we reflect upon our present situation, about what we are, about what we project, about our thoughts, feelings, and actions.

Each of us here has a psychology, and this is not a matter of believing or not believing, but observation.

Anger that leads us to madness exists within us.

Covetousness also exists within us, and we not only covet but moreover, there are some who boast about themselves as being saints, thus they covet to not be covetous.

Within us is lust that turns us into true beasts.

There is also envy within us, which has become the means of social action, because if we see that someone has a pretty, ultramodern, and flaming car, we envy him and desire to have a car like that, or even better. If we see that a friend of ours has bought a pretty house and also has a beautiful spouse, we envy our friend and desire to have a house better than our friend's.

And, if we want to boast about being virtuous, we affirm, "No, I do not covet, I am content with what I have: bread, shelter, and refuge, and that is all," even if the desire of conquering fame, honors, prestige, money, etc., is burning within us.

Pride is corroding our heart, thus each of us has our particular, individual pride. We love ourselves too much and that is very grave.

There are many sluggish gluttons—piles and piles—but we believe that we are not sluggish nor gluttonous, but "holy little saints."

The crude reality of the facts is that within us we have negative values that lead us to failure in these moments of world-wide crises and bankruptcy of all principles, in these precise moments in which the third world war approaches.

I say that each one of the psychological defects that we have in our interior is like a demon or a tenebrous entity. When one reads the four Gospels, we find a verse in which it is emphatically affirmed that the great Kabir, Jesus of Nazareth, the Christ, cast seven demons from the body of Mary Magdalene. Behold the seven capital sins. If they are multiplied by seven, and many thousands of sevens, then what the great Kabir threw from the body of Mary Magdalene was a legion.

Virgil, the great poet of Mantua, said:

*No, not if I had a hundred mouths, a
hundred tongues, and throats of brass,
inspired with iron lungs, I could not half
those my horrid crimes describe, nor half
the punishments those crimes have met.*

- *The Aeneid*, book six

Therefore, the Christic Gospel is correct
when affirming that each of us is a legion.

If we affirm in a clear and precise manner that the "I" is not something individual, but that it is a plurality, we would
not be exaggerating. Within each person
there exists the pluralized "I", i.e. I envy,
I love, I hate, I am afraid, I am lustful, I
have egotism, etc.

All of this multiplicity exists within us,
here and now.

We are speaking of the field of revolutionary psychology. We are affirming
that within us are multiple psychological entities, and this is already properly
documented. It is properly documented in
all the contradictions that we have in our
own mind. As soon we are affirming one
thing, soon we are denying it. Our mind
is like the weather. We are full of psychological acrobatics; we never keep the same

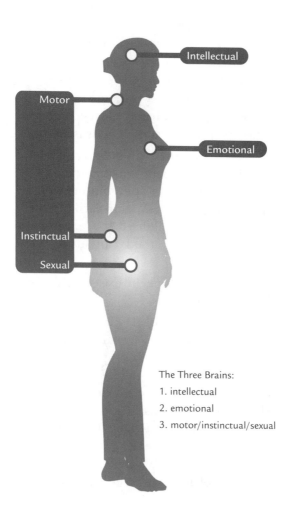

The Three Brains:

1. intellectual
2. emotional
3. motor/instinctual/sexual

opinion. Therefore, from where emerge so many psychological contradictions?

The brain is nothing but the instrument of the mind. The brain is not the mind. It is made in order to elaborate thought, but it is not thought. Therefore, let us delve thoroughly: from where do so many psychological contradictions come? Obviously, they come from the plurality of the "I."

If we state that each one of our "I's" has the three brains—namely the intellectual, emotional, and motor brains—we would not exaggerate. In other terms, we would say: in the same manner that the three-dimensional space of Euclid exists, likewise psychological spaces exist within each person. There is no doubt that the psychological multiplicity that flourishes in us is a reality within psychological space.

Nevertheless, ordinary physical senses are not capable of perceiving psychological space, but other senses can. The sense of psychological self-observation can perceive that space. Regrettably, the sense of psychological self-observation is atrophied. Yet, while we observe ourselves from moment to moment, we are able to develop that sense. When this occurs, the multiplicity of the "I" will be a reality for us; we will see it, and we will also intelligently perceive the psychological space.

So, each of us is a legion, and we have our consciousness extremely asleep. The intellectual humanoid is not capable of seeing, feeling, or touching the great realities of psychological space. We need to awaken our consciousness, because our consciousness is bottled up, inserted within all of those "I"s that in their whole constitute the self-willed, the myself, the itself.

We need to disintegrate those "I"s that personify our errors, and that is possible by means of psychological self-observation. It is in the field of practical life, i.e. in the factory, in the office, in the house, in the street, or in the market, or wherever we can discover ourselves. When in relation with people, the defects that we carry hidden spontaneously emerge, and if we are alert and vigilant as a watchman in the time of war, then we see them.

Once a defect is discovered, we must severely judge it by means of the superlative analysis of the Being. Any discovered defect must be studied and later disintegrated.

Obviously, the mind cannot radically alter any defect. The mind can justify this or that error, change it, pass it from one department to another of our understanding, justify it or condemn it, but never disintegrate it.

We need a power that is superior to the mind, a power capable of annihilating any defect. Fortunately, such a power is latent within the depth of the human anatomy. I am emphatically referring to the astral signature of fire. In a clear manner I am emphatically referring to God-Mother, the principle of love, the Eternal Feminine Divine. I am in a clear manner addressing the Divine Mother Kundalini Shakti, Stella Maris or the Virgin of the Sea, Tonantzin, Rhea, Mary, Cibeles, Adonia, Insoberta, Diana, etc.

God-Mother underlies in the depths of our own Being. It is a flaming power that only the Initiate with the sense of the psychological self-observation can perceive. Thus, if we appeal to that igneous and divine power—which is a variant of our

Athena (the Divine Mother) prepares Theseus (the consciousness) to battle the Minotaur (the animalistic ego).

own Being; God-Mother is our Being but a derivative—we can then totally disintegrate any psychological defect that we had previously comprehended in all the levels of our mind.

It would be enough to cry out, as when a child cries out to his mother because he is hungry or thirsty. Yes, it would be enough to beg to our Devi Kundalini Shakti for the disintegration of any previously comprehended "I"-defect; thus this is how it could be reduced to cosmic dust, to ashes, thus the consciousness that is bottled up within the "I"-defect would be liberated. By means of this procedure we could disintegrate all the "I"s-defects and liberate the totality of the superlative consciousness of the Being.

A liberated, emancipated consciousness is capable of seeing, touching, or feeling the great realities of the psychological space. A liberated consciousness is beyond the mind, and can perfectly discover the reality of all the phenomena that happen in the universe.

I want you to know that there are three types of mind. We can denominate the first one as "Sensual Mind," which develops its basic concepts via external sensory perceptions. This mind knows nothing about psychological space; it knows nothing about reality, God, etc.

Mr. Emmanuel Kant, the philosopher of Königsberg, wrote a book entitled *Critique of Pure Reason*. Kantian thought, with all its syllogisms, pro-syllogisms, quasi-syllogisms, etc., is formidable. With *Critique of Pure Reason,* Mr. Emmanuel Kant demonstrated to the world that the Sensual Mind cannot know anything about reality, about the truth, about God, etc., since it develops its basic concepts via external sensory perceptions. Therefore, it cannot know anything about the truth.

Now, there is a second type of mind. I am referring to the Intermediate Mind, where all dogmas, religious beliefs, etc. are deposited. Everyone is utterly free to believe in whatever they wish, therefore, we, Gnostics, would never pronounce ourselves whatsoever against other people's beliefs. We know how to respect our neighbor's religion and all religions, because we consider that all religions are like precious pearls linked on the golden thread of divinity. Nevertheless, religious beliefs are not direct perception of the truth, either. The Sun exists whether we believe in it or do not believe in it. The Earth will rotate around the Sun whether we believe it or not. The fire will burn our finger each time that we put it within the flame, whether we believe it or not. Therefore, what we believe or stop believing is not the truth.

Fortunately, there is a third mind, the Inner Mind, which is beyond the Intermediate Mind. If the Sensual Mind works based on its external sensory perceptions, then indeed, the Inner Mind works with the precise perceptions of the superlative and transcendental consciousness of the Being. Therefore, the awakened consciousness can know the phenomena of Nature in a direct, complete, integral, and unitotal manner, and thereafter transmit such data to the Inner Mind. Therefore, the Inner Mind knows about reality by means of the data transmitted by the superlative consciousness of the Being. Thus, the Inner Mind knows about the mysteries of life and death; it knows the origin of life. It discovers what the ignorant Sensual Mind cannot.

The Inner Mind knows from where we come, to where we go, and what the objective of existence is, etc. The Sensual Mind cannot know the phenomena of Nature in themselves. I.e. we see a flower, a carnation. The Sensual Mind says, "It is a carnation," but who told the Sensual Mind that this is the name of that flower? The Sensual Mind learned it at school, or it was taught to us at home or by other people. But, are we sure that this is the true name of this flower? This is how they taught it to us, okay, but what authority do they who put that name to this flower have? What is the true name of that flower? Are we perhaps the masters of the universal wisdom who know the name that the Divine Architect gave to this flower?

In the Inner Mind, everything changes. Thus, we say, "The true name of this flower is such-and-such, its components are such-and-such."

At school, at college, at the university, the chemical formula of this flower was delivered to us, thus we see in this flower the formula that they placed in our memory, but we are not seeing the flower, we are not seeing its true name, we are just seeing what they taught us, we are placing on the flower what we learned at school, at college, at the university, but we are not seeing *the flower*.

To *see it* is different. For this we must be open to the new, so that the flower can speak to us. If we want to know it, we must place ourselves in a receptive state. But, we are proud. We believe we are greater than the flower, thus we name it in this or that manner, and we say, "This is a carnation, and its chemical formula is this," because this is how they taught us at school. But we are not seeing *the flower*.

The consciousness can indeed see the flower and know its real name in the cosmos. The consciousness can know its true functionalism and its real elements. The consciousness can transmit that data to the Inner Mind, and the Inner Mind can comprehend it.

Regrettably, at the present time with our Sensual Mind, the only thing that we do is to project our ideas and concepts onto phenomena. With the Sensual Mind, nobody can learn the phenomena of the Nature and the cosmos, because life flows incessantly, and when we want to retain it—even for a moment—we kill it. Thus, only with the awakened consciousness expressed through the Inner Mind can we know for ourselves the phenomena in themselves, here and now!

There are two types of science: profane science and pure science. In pure science, theories do not exist, only facts. I.e. if I said to you that Count Saint Germaine, who lived in the fifteenth, sixteenth, seventeenth, eighteenth, nineteenth, etc., centuries, is still alive, you would believe that I am crazy, but nonetheless, I know count Saint Germaine and this is why I give testimony that he is alive. He lives based on a science that you do not know. This is the pure science, the science of the Superman, a science known by the extraterrestrials who travel throughout the infinite space, the science of the lords of life and death, the science of those who have opened the Inner Mind.

We are nothing but a part of the universal knowledge, and that is all. Yet, we can awaken our consciousness by destroying the undesirable elements that we carry within, thus transforming ourselves radically, so that we can become Supermen in the most complete sense of the word.

Now, at these moments of worldwide crises and bankruptcy of all principles, in these moments of terrible earthquakes and tsunamis, it is worthwhile that we explore ourselves. It is worthwhile that we try a psychological change, a radical transformation. It is worthwhile that we rise up in arms against all antiquated and extemporaneous concepts. It is worthwhile that we become true psychological revolutionaries, that we become true intelligent rebels capable of initiating a new civilization and a new culture...

Chapter 2
The Naked Truth

The following news appeared with a large headline on the front page of a very well-known newspaper in Mexico City: "Flying Saucers in France and United States Detected by Radar." Here we transcribed the text of this alarming news:

"Oklahoma City, Oklahoma. August 2nd (AP): Flying saucers reappeared last night in the Midwest of the United States.

"The highway police of Oklahoma indicated that radar at the military base of Tinker, near Oklahoma City, registered the presence of four unidentified flying objects that flew at an altitude of about seven thousand meters, but the military base refused to confirm or deny the news." [They hide it].

"On the other hand, three police patrols affirmed sighting the objects that flew in perfect formation for about thirty minutes. The color of these objects was first red, then transformed gradually into white and greenish blue.

"The sheriff's office in Wichita, Kansas announced that several unidentified flying objects were observed last night in the sky during some hours, at a height of some two to three thousand meters."

"Marmande, France. August 2nd (AP): A flying saucer was seen last night by a student near the city of Marmande in southwest France. According to the witness, it was an enormous luminous disc that landed in a cultivated field and soon took off and moved away at a vertiginous speed.

"From all the corners of the Earth arrives alarming news about unidentified flying objects. According to an eyewitness, one of those spaceships landed in France, and the crew of medium stature came from its interior. In that place, authorities found tracks of an unknown ship."

"In Argentina another ship landed on a mountain of difficult access. A farmer reported it to the authorities, who could observe the ship, but did not manage to reach it due to the steep landscape."

"In Australia, a cosmic ship flew over a control tower for space rockets at the moment in which the experts of the tower were following the trajectory of the rocket that photographed Mars."

The naked truth of this matter about flying saucers is that they really exist, and that they have been registered by radar and properly photographed. It is impossible for radar and photographic devices to hallucinate.

This subject-matter about flying saucers is already overwhelming, and even if intellectual-loafers and skeptics mock with irony those of us who affirm the existence of cosmic ships, whether they like it or not, the flying saucers are a concrete fact, properly registered by radar.

We are absolutely sure that intellectual-loafers do not like to face this thorny subject, due to that which is called self-love. Nobody likes for his self-love to be hurt, and since they love themselves too much, they are not willing to give up their beloved theories just because...

Intellectual-loafers think that human beings exist only on Earth. Their presumption is such that they firmly believe that only they have the right to live in this wonderful and infinite cosmos. Thus, this is how they are, and there is no way to convince them that they are mistaken.

Before the concrete facts, before the overwhelming news about flying saucers, the Gnostic movements remain firm, demanding that the "scientists" speak with frankness and stop hiding the truth about flying saucers or cosmic ships.

In the desert of Nevada, United States, the great North American scientist Adamski came into contact with Venusians who landed close to the place where he was making his investigations. This widely known scientist of world-wide prestige conversed with the Venusians.

In a South American country whose name we cannot mention exists a scientific society composed of ninety-eight wise men, disciples of Marconi. These wise people coexist with a group of Martians who regularly land in that region.

What bothers the intellectual-loafers the most is that this subject-matter does not become public, and that everything is made in secrecy. We ask the intellectual-loafers if they are so unconscious as to give a stick of dynamite to a three year old boy. What would happen to a boy who plays with a stick of dynamite? If these flying saucers were given to earthly humanity, we can be absolutely sure that the flying saucers would be utilized for war, and nobody upon the face of the Earth would be safe with their life. Let us remember the speed of these ships, their power of vertically ascending or descending, the power of apparently remain still in the air, etc. So, to give these ships to humanity would be like giving a stick of dynamite to a boy so he can play with it. Therefore, to the gentlemen intellectual-loafers who are abundantly displeased with the secrecy of this matter, we advise three things:

FIRST: to regenerate themselves

SECOND: a good dose of patience

Third: to abandon the mistaken concept of considering themselves the only inhabitants of the cosmos

The rocket that photographed Mars is not a wonder of science. From their terrible photographs taken from seventeen thousand kilometers away, it is impossible to retrieve information about whether life can or cannot exist on Mars. It is extremely stupid to deduce the vital reality of the planet Mars based upon a terrible photograph.

The innumerable craters on Mars do not signify that this planet is a dead world like the Moon. If the Earth were to be photographed from a distance of seventeen thousand kilometers, it is logical that such a photograph would be similar to that one which was obtained from Mars; in those photographs we would see something foggy, full of innumerable "craters." No cosmic photograph can inform us about the oxygen that a determined planet does or does not have.

Even though the gentlemen intellectual-loafers feel very annoyed and send against us all their defamatory dribble, the reality is that at different places on the Earth there already exist select groups of people who are in direct contact with the inhabitants of Mars, Mercury, Venus, etc.

Chapter 3
New York's Blackout

My friends, this night, we are going to discuss about a certain very fascinating article that I read in a magazine. This article is entitled, *"Doubt about New York's Blackout."* We will transcribe for you some paragraphs of the cited article:

"The greatest and most unexplainable electrical fault in history occurred exactly the ninth of November, 1965, at 5:28 P.M. On that date and time, twelve million New Yorkers suffered the consequences of a total interruption of the supply of their electrical energy. However, what the inhabitants of New York ignored was that in addition to their city, other cities were submerged in the darkness.

"Just make the emergency lights function, otherwise we are vulnerable to thieves.'

"The U.S. cities of Albany, Rochester, Boston, and Providence were in darkness, and Quebec and Ottawa of Canada also suffered from the same technical fault. Thus, across a broad area of 207,184 square kilometers, related with the provinces of Quebec, Ontario, and the states of Vermont, part of New Hampshire, Massachusetts, Rhode Island, Connecticut, New York, New Jersey, and Pennsylvania, no equipment that depended on electricity was functioning.

"Thirty-six million people—that is, more than the combined populations of Argentina, Bolivia, Chile, Paraguay, and Uruguay—were in darkness.

"Elevators, traffic lights, radios, televisions, and rotary presses ceased to function. Electrical ovens, electrical stoves, refrigerators, blenders, toasters, iron lungs, and even dentist millings became useless. Thus, frightful traffic jams were produced in the streets.

"The situation became even more critical because of the fact that the vehicles were left without gasoline, since the gas stations could not supply gasoline. The electrical pumps were not functioning and many cars were left abandoned.

"Almost a million people faced the impossibility of using the transportation in the metro stations.

"*I live about thirty miles away from here, and my son is sick. We cannot move.*'

"The city was congested with the blackout. The airplanes that were going to

land or take off were having difficulties to
do so when the illumination on the run-
ways was momentarily off.

"*'It is urgent for us to land, because we are almost out of fuel.'*

"Anarchy was increasing. There were
dead because of many accidents. Thieves
abounded. People were distressed because
of the darkness. The situation was so
critical and unexpected that the gather-
ings that were being attended in the crys-
tal palace of the UN in New York had to
be suspended.

"As the duration of the blackout was pro-
longed, the highest authorities were becoming aware that they were confronting
a very dangerous alteration of civic order.
*'Last hours' reports indicate that panic and
disorder are increasing in the streets. This situation is chaotic! Only buildings like this one,
with their own electric plants, are the ones with
electric light. Yet these are very few!*

"*'Nevertheless, what is indeed the most critical matter, is that in addition to the panic and
chaos of this moment, in this immense zone of
207,184 square kilometers... the radar systems
do not function! Therefore, any aerial projec-
tile can cross our sky in these precise moments
and we cannot detect its presence! Moreover,*

we cannot send even one of our guided projectiles in order to intercept it, because the buttons with which we command them depend on the electricity for their functioning. It is useless; this immense device is good for nothing!

"'Gentlemen, we are one of the most powerful countries of the Earth. Every year we expend million of dollars in armaments and in fortifying our system of security. Nonetheless, facing a situation like the present one, we are impotent.

"'A civilization as technical and powerful as our own has an Achilles' heel—that is to say, a great weakness: the electricity!

"'Have you asked yourselves, what good is radar to us, as well as guided projectiles, the telegraph, the telephone, and television? Good for nothing! Absolutely good for nothing!

"'Of course, the General is right, since we cannot transmit an order! Or receive any information either!

"'This is the bankruptcy of our military and industrial power! Everything is paralyzed!

"'Gentlemen, please have a little more composure... it is good to remind you that the government to which all of us belong is totally aware of the importance of electricity.

"'In each electric plant, great electronic cerebrums and computers exist, which are in charge of investigating every centimeter of

all the cables which conduct the electricity.
Thus, when one of these electronic cerebrums
discovers for instance, that one of the cables is
overcharged with electricity and that a danger
for being fused exists... ('Line number five is in
danger!')... then automatically, that charge is
distributed among other cables of the same system... ('The overcharged force was distributed
between the lines seven and eight'). Likewise,
the computers indicate when a great fuse is
burned, to which line it belongs, and the exact
place of the problem. ('In the third cable of the
system the Albany system, the problem is located between tower 17-B and tower 18-B. We
are leaving immediately to go there').

"'Moreover, if in a zone a problem could be
produced that cannot be caught by the electronic central plants, then that zone is isolated from
the rest, so that it does not cause any damage
to the other systems of electric induction. ('We
have isolated the south zone of Montgomery
and are proceeding to make the repair').

Each elevator, each fuse, each switch, and each
connection are carefully checked every moment. ('I am sure that from one moment to the
other, the problem that occurred will be fixed
and we will have electricity again. The blackout has lasted only one hour').

Disgracefully, those were optimistic calculations tat were not confirmed by the crude reality.

"*My son, my son; I could not see him in this obscurity, he left, running towards the alley!*"

The great blackout endured exactly twelve hours, not even a decimal of second more!

"*Frank, come on, we have light again!*"

"The most precise electronic chronometers did not need to be touched in their needles in order to be fixed, because when they received the electric charge anew they again indicated the hour with great exactness. "*It is amazing: it is showing the right hour!*"

"*We must demand an explanation!*"

"Thus, in the same mysterious way that the electricity was interrupted, the electricity started to run again in each cable. Thus, a blackout of that unexpected duration like the former and which "curiously" had the exact duration of twelve hours, provoked uncertainty in a great sector of North America.

'*Let responsibilities be determined!*'

"So, common and current citizens, industrial people and journalists, asked the government of the United States for an explanation about this phenomena.

"'More telegrams have arrived, sir. They demand for us to clarify what happened.'"

There is no doubt that a few extraterrestrial human beings were capable on their own to paralyze the whole electrical system of New York and many other cities, as already stated.

We see then the fundamental difference between the human being and the intellectual animal.

It is obvious that everyone considers themselves to be within the "homo sapien" or human kingdom, to be more specific. The philosopher Diogenes walked with his illuminated lantern through the streets of Athens in search of a man (a real human being), yet he did not find one. Pilate in the Gospel, when presenting Christ to the multitudes, exclaimed, *"Ecce Homo:* behold the Man."

How difficult it is to attain the level of human being! Nevertheless, everyone believes themselves to be human.

The human being is the true king of creation and he can perform marvels, great marvels such as that of the blackout of New York.

We are absolutely sure that half a dozen real human beings could paralyze not only the electrical system of the United States, but moreover, all of the activities of the whole world. Half a dozen authentic, legitimate, and true human beings could take the whole planet Earth for themselves without shooting a single bullet, and in a matter of minutes.

Question: How can it be possible that a half a dozen human beings could achieve that?

Answer: Oh, respectable young lady, you are overwhelmed with my words! Well, it is necessary for you to comprehend that the authentic human being is a king of creation. Only one human being could, in a matter of seconds, disintegrate the planet Earth, to convert it into fragments that would rotate around the Sun.

Question: Can that sort of human being perform that much without being Self-realized?

Answer: When are you going to understand me, respectable lady? Please comprehend that it is not possible for a human being to exist without Self-realizing himself. What happens is that you think in a mistaken way. You suppose that these intellectual animals who populate the face of the Earth are humans. Behold, this is your error.

Nonetheless, inside the intellectual animal exist surprising possibilities. The rational homunculi commonly called "human being" is a chrysalis within which the human being can be formed. Have you understood me?

Question: What can we do in order for the human being to be formed within that chrysalis?

Answer: Oh, respectable young lady, in this basic book, I cannot give you those explanations. Come to our Gnostic studies, study our books: read *The Perfect Matrimony, The Mystery of the Golden Blossom, The Three Mountains, Parsifal Unveiled,* etc.

Question: Then, in accordance with what you are stating, are true human beings dangerous?

Answer: Listen to me, noble lady: the legitimate human beings, in the most complete sense of the word, are not dangerous, as you apparently think that they may be. If they wanted to make the planet Earth explode into pieces, they would have done it already. If they wanted to invade and enslave us, they could have done it already, many centuries ago. Nonetheless, the authentic human beings neither assassinate nor enslave; they do not invade nations or worlds.

What happens in this day and age is that they are visiting us in order to help us, because they are compassionate. They will assist us in the supreme hour of the great cataclysm. After the tremendous catastrophe that is expected for us, and when the Earth has renewed conditions to support life, then they will establish a new civilization and a new culture on this planet Earth.

The intellectual animals are puzzled by the fact that these extraterrestrial beings do not fit all within our environment and that they do not submit themselves to our obsolete and degenerated order of things.

Question: Then, according with what you stated, true human beings do not exist on the planet Earth. Are all of them extraterrestrial?

Answer: Respectable young lady, listen to me, human beings also exist on the planet Earth. Nonetheless, they must be sought with the lantern of Diogenes, because they are very difficult to find. Nevertheless, you can see that everybody boasts of himself as being a human being.

Question: What is the objective for us to become authentic human beings?

Answer: Listen to me, respectable young lady. The butterfly, which joyfully flies in the sunlight, emerges from the chrysalis. Likewise, from the humanoid chrysalis, the authentic, legitimate, and true human being can emerge. This is the original purpose of divinity. Would you be against the purpose of God? God wants each of us to be converted into a king of creation. Only thus can we enter into the kingdom of Melchisedeck. Only thus can we enter into paradise. This is why Jesus Christ came into the world: He came to help us. He wanted each of us to become a king of Nature. Now, you will comprehend the reason for which we have to be preoccupied.

Question: What supposedly produced the blackout?

Answer: With much gladness, I will answer the question of the gentleman who is listening to me. It was completely and officially proved that the New York blackout was produced by an extraterrestrial cosmic ship. Based on this event, the government of the United States created a scientific department with the evident purpose of investigating this matter about the UFO's.

It is necessary to remember that moments before the blackout, two fighter planes from the United States—which were armed with high power projectiles—were in pursuit of two UFO's. One of the UFOs became lost within the infinite space, while the other one descended over the electrical energy plant of Syracuse. Thereafter, the blackout happened, and this is a completely proved record. Those airplanes, armed with rockets, were good for nothing; the best weapons of the United States proved useless.

Question: What was the motive for the blackout, and what was the message?

Answer: I will answer the gentleman. Listen to me, if they wanted to reduce New York to ashes, they would have done it in a matter of seconds; however, they are not perverse. They look at this race of intellectual animals with infinite compassion.

They were pursued by fighter planes. The U.S.A. wanted to destroy them. They indeed were not greeted with a beautiful welcome. They were not approached as brothers. Nonetheless, the only thing that they did to the intellectual animal was to demonstrate the state of unconsciousness and weakness in which he is situated. Thus, when paralyzing the whole electrical system of New York, they gave a demonstration, trying to make the rational beasts comprehend the shameful state in which they are situated.

At the present time, after such an event, the rational animals should be studying themselves, eliminating their passions and vices, repented, purifying, and sanctifying themselves.

Question: Do you consider that the extraterrestrial beings think that we understood ?

Answer: Respectable friend, it is obvious that this kingdom of rational animals have their consciousness asleep, therefore, for such reason, they are very far from comprehending the teaching that was given to them.

The rational animals are not human beings. Nevertheless, they believe themselves to be omnipotent, powerful, super-civilized, and super-transcended, etc.

Question: Could you explain to us scientifically what they did in order to produce the blackout?

Answer: Oh, respectable sir, this is a matter belonging to a superior type of electrical science. I firmly believe that it is possible to deviate the electricity by orienting the current in a different way, to alter the polarities, to invert the charges; then, it is clear that any city would be in darkness. To study this matter in a very detailed way and thereafter publish the knowledge of it would be to put in the hands of rational beasts the weapons of the human being. What would the rational beasts do with such knowledge? Place your hand over your heart and answer the question. Understood?

Chapter 4

Flying Saucers and Little Green People

From Lima, Peru came news dated August 2nd that stated:

"A flying saucer with its crew, dwarves of greenish color, was seen by a young student last night on the roof of a house of this capital, according to statements that he gave today to 'The Commerce' newspaper.

"This visit is in addition to the one reported last week by a guardian of Chosica District, about forty kilometers from Lima, who claimed to see on the patio of a factory a flying saucer protruding a tube similar to an elephant trunk, which disappeared after ten minutes of observation.

"In regard to the flying saucer of last night, Alberto San Roman Núñez—of about fifteen years of age—affirmed to have seen a wrinkled greenish being of about ninety centimeters in stature [three feet] that slid on the roof.

"Shortly after, the ship projected a reddish light, in the middle of which it took off and flew, leaving tracks on the ground in which four landing marks are visible."

Green skin color may surprise many people, but we Earthlings have races of black and yellow and red skinned people that could surprise the cosmic visitors.

Indeed, not one of the eyewitnesses of flying saucers and extraterrestrial crews would dare to assert that these mysterious visitors have features that differ from those of us, the wretched Earthlings.

It is lamentable that science fiction has been dedicated to propagate false ideas or fantasies about the figure and form of the extraterrestrial visitors.

It is clear that skin color varies according to climate, atmosphere, etc., nevertheless, the human form—whether gigantic, medium, or small—is always the same.

Science fiction has been in charge of propagating everywhere—either through the radio, or through cinema or by means of television—tremendous lies that are detrimental for humanity.

Defamatory calumnies against the extraterrestrial visitors have arisen, since the minds of Earthlings judge in accordance with their perversities, thus wanting to see in our noble visitors all the hatred of the minds of Earthlings, i.e. all the atrocities of Hitler, all the monstrosities of the inventor of the hydrogen bomb, all the bloody purges of Stalin, etc.

Perverse Earthlings do not want to recognize the noble intention of our extraterrestrial friends. If they wanted to take the planet Earth and to enslave all of its inhabitants they could do it in matter of minutes, because they have sufficient elements to do it. If they wanted to destroy us, they would have already had done it, because they have atomic and scientific instruments with which they can scatter into pieces any planet in space.

Let us remember that long before we Earthlings knew mathematics, they were already navigating sidereal space.

Our extraterrestrial friends know the planet Earth better than us, and they do not have any interest in enslaving or destroying us, as "mysteriously" propagated by the science fiction of these times of rock 'n' roll and rebels without a cause.

Our extraterrestrial friends know the critical hour in which we live, and only want to help us. We need their aid with extreme urgency, because we Earthlings have totally failed.

If the barbarian hordes of Earthlings continue with their stupid intention—that is, to capture or to destroy the cosmic ships that visit us—then lamentably we will lose the shining opportunity that our brothers from space are offering to us.

They want to establish personal contact with us, but instead of welcoming them with true respect and love, instead of offering them hospitality, we send fighter jets to intercept them. Everybody wants to destroy them. Indeed, we are behaving like savages, very far from all type of civilization and culture.

The hour has arrived to change our militant attitude and to offer our friendship and affection to our brother visitors from space, since they come to help us, not to destroy us.

We, the Gnostic brethren, must begin to give the example by establishing friendly signals—circles with a point in the center—upon our properties and the roofs of our houses of our countries.

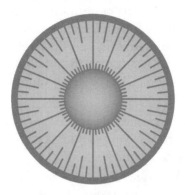

Lines radiate from the central point, which go towards the periphery; from the periphery other small lines radiate that, although they do not arrive at the center, nonetheless indicate they are going towards the central point.

Draw the mentioned central point of the circle and paint it with a beautiful golden color to symbolize divinity. The lines that radiate from the periphery to the central point can be blue and short, in great quantity. The lines that radiate from center towards the circle are clearly connecting the central point with the circumference, and these can also be blue.

This is the symbol of divinity in Martian religion. We can use it by placing it upon the roofs of our houses, on our lands, doing it with luminous centers or simply painted, this, in order to establish friendly relations with the inhabitants of Mars and all of the inhabitants of the cosmos.

Such a symbol means that everything comes from Divinity and everything returns to Divinity.

Use this symbol in order to offer friendship to the inhabitants of space, even if intellectual-loafers laugh at us. All of you already know that the intellectual-loafers are one hundred percent skeptical, even though they boast of being super-civilized. They believe themselves to be very wise when using satire and fine irony against all of us and anyone who does not want to think like them.

Chapter 5

Pure Science

The Lemurians of ancient times had developed objective reasoning. Likewise, many people in Atlantis possessed that type of reasoning. Obviously, the human beings from the Polar and the Hyperborean epochs also possessed objective reasoning. Yet, it is regrettable that in this current age of Kali Yuga very few have developed objective reasoning within their interior nature, since subjective rationalism is in fashion; it is what predominates in these times.

Subjective rationalism is the basis for Kalkian personalities. Let us understand by "Kalkian personalities" those pseudo-esotericist, pseudo-occultist, and pseudo-scientist people from this modern epoch, along with all of their know-it-all types of foolishness. Never has there been as much darkness as there is in this epoch of Kali Yuga.

In fact, only those few who have developed objective reasoning have access to pure science.

Thus far, approximately two-thirds of all spacecraft destined for Mars have failed. As one example, the failed Mars Climate Orbiter, depicted here at launch, was a loss of $327.6 million dollars.

Let us distinguish between the pseudo-science of this epoch of Kali Yuga and pure science.

Kalkian personalities—the know-it-alls of the Tower of Babel, the studious igno-ramuses of subjective rationalism—will never have access to the pure science.

As an example of what pure science is, in complete opposition to the ultramod-ern pseudo-science, let us observe the fol-lowing examples: the scientists from the present "Tower of Babel" (NASA) launch rockets—shoddy scrap-heaps—into outer space, propelled by volatile combustion. Thus, after performing their circus feats, they finally land their famous so-called astronauts on the moon. Behold here the outcome of the merely subjective rational-ism.

In opposition and as an example of pure science, there are interplanetary ships propelled by solar energy—within which liquid combustible is not neces-sary—ships that travel from galaxy to gal-axy at velocities faster than the speed of light. The circus adventures of the famous astronauts nor anything of the sort are needed, since this belongs—as I already stated—to pure science, to objective reasoning.

Looking at this, therefore, in a completely logic confrontation, we see on the side of subjective rationalism the circus-like rockets, and on the side of pure science and objective reasoning are the extraterrestrial ships.

The skeptical and incredulous ones smile abundantly when we talk about extraterrestrial ships that travel from galaxy to galaxy, nonetheless, a doctor—a celebrity of NASA—was taken in one of those ships and wrote a book that is in circulation. Therefore, what we are stating here has complete confirmation.

Chapter 6
Flying Spheres

Extraterrestrial ships are not "flying saucers" but "flying spheres." These metallic spheres mimic planetary rotation. Their rotating motion is from left to right, and when slow, makes them ascend or descend.

In the center of the sphere there is something that in shape and function resembles fins or trowels in the form of fan, that absorbs the cosmic ether, which, conducted by a tube, is then burned by means of a ray of light (this is not similar to the spark ignition system that we use). By clairvoyant observation, one sees that it resembles a small wheel similar to emery wheels that we use in order to sharpen steel objects. The ray that burns the ether is produced from this small stone that turns incessantly. This produces the inexhaustible fuel that drives the sphere.

The burning of the cosmic ether and its jet expulsion brings about two continuous movements of the ship: one revolving and one propelling. In addition, the jet expulsion has a third use: it serves as a rudder for the ship. With a slight downward movement the sphere begins to ascend; with the upward movement of the tail rudder the sphere inclines downward and descends. Likewise, in order to turn left or right, the jet expulsion allows the sphere to be maneuvered.

So, the cosmic ether—conducted by a tube, burned, and producing the external rotation of the sphere—also warms it, producing enough heat to protect the small pilots, thus providing the capability of traveling amidst great planetary cold without the accumulation of ice or other gases that are found in the firmament.

The sphere rotates upon its axis, and the cabin where the small pilots are accommodated is fixed to this axis, therefore it is stable regardless of rapid external maneuvers.

The cabin contains two small dials: one indicates direction and the other speed. They adjust the jet expulsion by means of a small handle, and the speed by means of a button or accelerator that they control with their foot; this accelerator goes

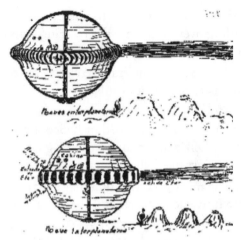

A drawing from the original edition of the book.

in and out and can be maintained on a certain point in order to keep consistent speed.

In order to land on Earth, a fan opens underneath. This fan picks up from the landing field a very fine soil in the shape of a cone, and smoothly lands upon it. The resumption of the rotating motion is enough in order for it to quickly take off; when on land the sphere maintains a reserve of pressurized ether that brings about the movement when it takes off. Until that moment, the rotating movement is what prevents the sphere from falling over; otherwise the rotating ship would crash against the ground.

The speed of the craft is relative to the intensity of the jet expulsion.

Movement from right to left is similar to what is accomplished by helicopters; the sphere can remain fixed at a given point just like helicopters.

The most powerful fuel that exists in the universe is the ether. Ether is found everywhere. Therefore, these cosmic ships have discovered the secret of perpetual motion that our wise people have looked for. When the sphere turns upon its axis, it introduces the ether; when the ether is burned, it produces force, and with its jet expulsion it continuously gives move-ment to the sphere, or rather, it rotates and impels it at the same time, thus, this movement is continuous.

This indicates that airplanes propelled by gasoline will soon be obsolete, since gasoline as a fuel is very heavy, dangerous, and occupies much space; all of this will pass and will be in the museums of antiq-uities. Therefore, the little wings will soon be unfashionable; we will need to fly like the heavenly spheres.

The rotating motion, as we said, is controllable—that is to say, it increases or decreases according to the intensity of the jet expulsion.

There are small and big spheres, capable of transporting enormous crews.

The inhabitants of our planet Earth are not capable of traveling to other planets. However, the "men-children" from other planets will come to civilize us, because we are savages.

In order to conclude this study of "flying spheres," I must say to the men of science that this is not a more or less pompous nor more or less vague theory. This is a palpitating reality.

The four collaborators who investigated these ships were inside of them and conversed with the small pilots, therefore, this study is neither a theory nor a hypothesis, nor is it an opinion. It is simply a reality. The scientists can perform the same experiment if they learn how to project themselves in their Astral Bodies.

In our world, many people believe that "flying saucers" are instruments of destruction or secret weapons invented by Russia, and that those machines are possibly moved by radar or other similar inventions, but there is nothing of the sort. These are not machines of destruction like those invented by Earthlings. These are perfect ships that will bring civilization to us, because we are not civilized: we are anthropophagi with tuxedos.

In Russia, there is a wise person who by means of radio waves has managed to communicate with these pilots. This wise person is helped by Uan Weor, the Weor of Russia, who is one of the seven Weors of the Holy Gnostic Church. Our brother Sir Weor of the United States does not ignore this either.

The scientists who want to investigate the flying spheres must study this book minutely. Let them work on the development of their own internal or occult powers, by means of which they will triumphantly and victoriously be able to enter into the amphitheater of cosmic science.

When the pilots of those flying spheres settle their dwelling on Earth, they will illuminate us, and then indeed there will be splendors and wisdom on the Earth. They are wise and holy at the same time. They are small in stature and have rosy faces like the aurora. Their mission will be to illuminate the future humanity of Aquarius.

The report given by Mr. Lapides, affirming that the "flying saucers" are a secret weapon of the United States is not true; these are mere suppositions or news in order to flatter certain readers. Also the information provided by Mr. Echeverria Márquez based on an English document of a British official's property wanting to explain the mystery of the flying saucers is also a simple supposition. All of this information is just trying to claim an invention that is not from our planet, and in order to provide flattering news to thousands of readers.

If the flying saucers were secret weapons, why would they fly over cities of little importance, over the fields, and avoid encounters with other airplanes, causing unexpected movements among them, flying calmly and in short, in view of thousands of spectators?

Like it or not, the stubborn ones of the twentieth century are going to have to be convinced that the flying saucers are ships from other planets and piloted by human beings million times wiser than the jack-asses with frockcoats, eyeglasses, and glass tumblers of our heartbroken Earth.

Chapter 7

Cosmic Ships

It was around the year 1950 when we spoke for the first time about flying discs [the previous chapter]. In that year we emphatically affirmed that such flying discs are indeed spaceships maneuvered by inhabitants of other planets.

In that year many laughed at our affirmations, yet now many events have provided evidence to support us, since in the United States there is a scientific department solely dedicated to the investigation of these spaceships.

The law of accidents also affects those ships; thus, several have clashed or violently exploded in the air. The United States has in their hands fragments of some of those ships.

In this book we do not intend to demonstrate the reality of the interplanetary ships, since that reality is already totally demonstrated. Now we only want to extend the information that we gave to humanity in the year 1950 in the first edition of our book *The Perfect Matrimony*.

The spaceships have their history and their traditions. Indeed those ships were created by Angels, Archangels, Seraphim, etc., who have their own bodies of flesh and bone.

Many cosmic traditions mention Saint Venoma and his cosmic navigation system. Saint Venoma is an Angel with a body of flesh and bone. Saint Venoma was born on the planet Soort where he dedicated himself to investigate the Law of Falling. Lo and behold, beloved reader, the formulation that Saint Venoma gave to this cosmic law:

> Everything existing in the world falls to the bottom. The bottom for any part of the universe is its nearest "stability." This "stability" is the place or point upon which all the lines of force arriving from all directions converge.

> The centers of all suns and of all the planets of our universe are such points of "stability." They are the lowest points of those regions of space upon which forces from all directions of that part of the universe definitely tend towards, and where

they are concentrated. In these points is also concentrated the equilibrium that enables suns and planets to maintain their position.

In this formulation of his, Saint Venoma said further:

> Everything dropped into space, wherever it may be, tends to fall on one sun or another, or on one planet or another, in accordance to the sun or planet belonging to that part of space. Each sun or planet is the "stability" or bottom for that space.

Having verified his deep investigations, Saint Venoma knew how to use that cosmic particularity for the locomotion of spaceships. The spaceships designed by Saint Venoma were based on the principle of the Law of Falling.

The only serious problem was related to planetary atmospheres, since they do not allow objects in space to fall straight. Saint Venoma managed to solve that problem easily, thus constructing wonderful spaceships. We do not intend in this book to thoroughly explain all the mechanisms of those ships because, besides being too complicated, it would be too tiring for the reader.

Under the direction of Archangel Adossia, the inspection approved and blessed the works of Saint Venoma.

The ships of Saint Venoma were moved by the magnetic force of the planets, and were very fast. Nevertheless, the most serious problem appeared when the ships approached any planet or sun in space, since a great deal of complicated maneuvering was necessary to avoid a catastrophe. Few were the Human-Angels who could drive those ships. It was very difficult to steer the ships of Saint Venoma, thus, everyday great demands were made upon the technicians who had to drive these ships. Nevertheless, the system of Saint Venoma was a technical revolution for its time, and gradually completely displaced all the previous systems.

After many years of cosmic activity, the system of Saint Venoma was entirely superseded by the revolutionary system of the Archangel Hariton. This Archangel is a true Man in the most complete sense of the word, and has a body of flesh and bones like any person.

The wonderful experiments of Mr. Hariton began under the supervision of a great wise man known in all the cosmos with the name of Adossia. This wise man is also a gentleman who already acquired the Archangel degree.

Modern spaceships are based on the works of Archangel Hariton. All the technical functionalism of these ships is performed upon the basis of perpetual motion. However, since this it is not a mechanical text and we are not mechanics either, we abstain from describing the mechanism of these spaceships.

Millions of spaceships—as numerous as sands of the sea—travel throughout the infinite outer space. Cargo spaceships are gigantic and carry within their gigantic bellies small spaceships that they use in order to descend to the planets. This is similar to the great cargo boats that carry on board small boats that are used in order to disembark on the shores.

Any mature humanity of the cosmos has the absolute right to receive these spaceships. Normally, the elder brothers help the younger ones, thus when a given humanity reaches an age similar to ours, they receive a visit from other planetary humanities. Then the elder brothers initiate the younger ones in cosmic travel, give them some of their spaceships, and teach how to construct them.

In the times of Atlantis, spaceships normally landed in the airports of the city of Samlios. At that time, the inhabitants of other planets visited the Atlantean kings and coexisted with them in their palaces.

When humanity became morally corrupted, then the elder brother humanities from other planets stopped visiting us.

In these times of the bankruptcy of all spiritual values, we are again to be helped by our elder brothers from other planets. Now, we need extra help, because we have failed and are on the brink of a great cataclysm.

This terrestrial humanity has reached the maximum of corruption, thus the aid of our elder brethren is urgently needed. Already, several inhabitants of the Earth have been taken on a promenade to other planets of the infinite space.

In the Republic of Mexico, two men were taken to Venus; one is a resident of Jalisco and the other lives in the capital city of Mexico. In the next chapter we will talk about the latter. We know that in Brazil is another gentleman who was taken to Mars. There is not the slightest doubt that all of us can visit other planets of outer space.

We are on the eve of a great cosmic cataclysm and will be warned before the great catastrophe.

The elder brother humanities from other planets will try to save us and will officially enter our major cities to announce the danger that awaits us. However, if we continue with atomic explosions and vices and all types of perversities and wars, the catastrophe will be inevitable.

It is good to know that a group of Tibetan Lamas already have some of those spaceships. They received them from our brothers from other planets and have them very well stored in certain secret places in the Himalayas.

Some citizens from other planets already walk the streets of our cities; they dress as any fellow countryman and nobody recognizes them. They study our languages, traditions, and customs, in order to help us.

We are going to be helped on a great scale. We urgently need such help, because we have totally failed.

Many spaceships now land in the jungles of Brazil, southern Argentina, etc. and in other places where they have secret airports. Some of their cosmic crew remain with us.

Those who think that the visitors from those humanities from other planets are trying to destroy us are mistaken. Those who suspect perversity from our visitor brethren are mistaken.

It is clear that they have weapons with which they can paralyze men and machines. It is logical that they are invulnerable because they are very well armed and protected.

If they wanted to dominate this planet, they could do it in seconds, since they have special weapons in order to do it. Likewise, they could destroy this planet, to scatter it into pieces, but indeed this is not what they want. They are not destructive, they respect life. They are not like the perverse Earthlings.

Our brethren visitors only want to help us. Thus, each and every one of us, inhabitants of the Earth, must prepare to receive them. At this moment, inhabitants of Mars, Venus, Mercury, etc., live in all the great cities of the world. They study our languages and customs in order to help us.

Spaceships land in isolated places and sometimes they leave some of their crew from other planets, who dress as anyone else walking on the streets of New York, Paris, London, etc. without anybody knowing about them, since they are very similar to us in their physical appearance, even though many of them are much more beautiful and perfect.

The fantasies of many authors who imagine that the inhabitants of other planets have different features from the human beings of the Earth are absurd. The physical features and forms of the physical body of all humans of the cosmos are always similar.

Venus

Chapter 8

A Mexican Traveled to Venus

Here in Mexico City we know a man who traveled to the planet Venus. We had the high honor of personally visiting and meeting Salvador Villanueva Medina. Salvador is not at all fictional or unbalanced. Salvador has been examined by psychiatrists and they have reached the conclusion that he is a mentally normal, balanced man. Neither does Salvador does make a living from his extraordinary adventure or on the book that he wrote, entitled, *I Was on the Planet Venus.*

One winter night we arrived at the door of his house. We were lucky enough to have been welcomed by him. His family was watching TV, yet in a very amiable manner they turned off the television and left us alone with him in their living room.

This gentleman is a mechanic by profession. He fixes cars. This is what he does for a living. We ourselves have been in his body shop watching him work. He is practical, one hundred percent. We keep the address of this gentleman private, since we do not have authorization to give it to the public.

He is a very sincere and kind man. He is not an occultist or a spiritualist or anything of the sort. He does not boast of being a wise person. And, in spite of having lived a most extraordinary cosmic adventure, indeed he is not proud about it.

We limit ourselves to two things: first, to give testimony that he is an absolutely prudent man dedicated to his work and his family. Second, that indeed this man happened to experience a formidable adventure, but he does not make a living from it.

Salvador Villanueva Medina narrated what happened to him, and this has brought him many sufferings, because, as always, the loafers, the skeptics, the stupid idiots have made a mockery of him.

Beyond any doubt, Salvador was on Venus, and he accomplished his duty of informing this to his fellow men, although they make fun of him.

"Whosoever laughs at what he does not know is an ignoramus who walks on the path of idiocy."

We do not intend in this book to narrate in detail what happened to this man; we only want to talk about it in synthesis, and that is all.

In the month of August of the year 1953, this man was on the planet Venus. His name is Salvador Villanueva Medina. Said event happened unexpectedly. On the highway of Laredo in Mexico, he was driving a rental car taking an American couple to the United States.

He drove about 484 kilometers when the car broke down. The "gringos" left the car and went in search of a tow-truck to take the car to the nearest town in order to repair it.

This is when his adventure began. Salvador went underneath the car to try to repair it; suddenly he heard some steps on the sandy roadside and somebody asked him in perfect Spanish what happened to his car. Salvador kept silent, and when he came out from underneath the car, he found himself before a strangely dressed man whose stature measured about one meter and more or less twenty centimeters [about 3.5 feet].

The body of this man was of an extraordinary perfection, his skin white like the ermine, and his entire appearance was filled with beauty. What called Salvador's attention more were the unusual uniform and mysterious shining belt with perforations from which strange lights shone.

The man had long hair and wore a very special metallic helmet.

The words spoken between them at that moment were few. Salvador limited himself to ask this mysterious personage if he was a pilot. The personage answered that his airplane—"as we Earthlings name our flying ships"—was a short distance away. Thus, after saying these words, the strange personage courteously left and soon he went towards the mountain. Salvador told us that after this event he decided to sleep calmly within his car...

The most interesting moment happened later, when Salvador was sleeping: strong knocks on the window of the car surprisingly awakened him. Without thinking too much about it, Salvador opened the door of his car; then, his surprise became astonishment when he saw the same personage accompanied by another who had the same aspect and the same suit. Salvador invited them to enter his car. Once inside, when stretching his right arm over them with the intention of helping to close the car door, he momentarily felt an electrical current that paralyzed his arm.

The conversation inside the car was wonderful. They declared to Salvador that they came from the planet Venus and gave a lot of information about this planet. They said that on Venus the highways endlessly extend and with many crossing arterial roads which are immediately below their highway levels in order to avoid accidents. On Venus the vehicles do not consume vegetal or mineral fuels because they are detrimental for the organisms. The Venusians use solar energy to power their vehicles.

At first, Salvador did not believe them, and even became offended thinking that these gentlemen were making fun of him. Thus, Salvador even said to them that only the planet Earth had inhabitants, and that he had learned this through the affirmations of the wise men from the Earth, etc. They then asked Salvador the following questions: "What makes them to think in such a manner? Perhaps the deficient methods they utilize in order to make their calculations? It is not perhaps too much pretension to think that they are the only beings who populate the universe?"

These questions were already very unusual for Salvador, and in addition the color of those very white faces, their expressive eyes, their strange voice, their strange helmets, their mysterious belts, etc. made him think very much.

To narrate the conversation that Salvador had with those Venusians would be too long; in short, they told him how life was on the planet Venus, how they lived, what they ate, how their cities were, their streets, etc.

They also removed from him all doubts by explaining to him that they could turn a detrimental thing into a beneficial thing, and artificially make their climate his atmosphere, etc. So, with those conditions, if Venus were uninhabitable, they could make it inhabitable because their scientific advances allow them to do it. However, it is clear that Venus is perfectly inhabitable.

Those Venusians informed Salvador widely about life on Venus. They dissipated his doubts by explaining to him that on Venus by means of special scientific systems they created an artificial, benign, and uniform climate, thus turning their world into a delectable dwelling.

Dawn was arriving, so in a very amiable manner the Venusians invited Salvador to accompany them to the planet Venus. Thus, Salvador left the car and followed these mysterious men. Certainly, after a while walking among the mountains, Salvador paused before the majestic spaceship.

It was a majestic and imposing flattened sphere that leaned upon three supports that formed a triangle. Salvador stated that the construction of the ship was impressive and gave the impression of being a great fort.

Salvador entered the ship; the doors were closed and they left for the planet Venus.

Everything that Salvador saw in Venus was extraordinary. The Venusian civilization is formidable.

On Venus, civilization has arrived at its peak. There, money is not needed. Each citizen works two hours daily, and in exchange one has the right to everything a human being needs for life, namely: transport, food, clothes, vacations, science, etc. Everything belongs to everybody. If one needs a car, one takes it, drives it, and leaves it in its parking lot. If one is hungry, one eats in any hotel and pays nothing for it, because since everybody works, everybody has the right to everything. If one needs to dress, one asks for a suit in a store and does not pay anything for it, because since everybody works, everybody has the right to get dressed, etc.

On the planet Venus, cars are powered by solar energy. On their world they have a single sea, but it is three times deeper than ours. Their main nourishment is taken from that sea. So, they take from the sea all the necessary elements to build their buildings, make clothes, vehicles, and sixty percent or more of their food. Their orchards are upon the roofs of their houses and buildings. They said that their ships can either be in the air or in the water, and that in the deep sea are gigantic factories, which are in charge of scientifically selecting and taking advantage of the fish for their nutrition. So, on Venus fish and fruits constitute their basic food. On Venus there are no governments, nor mother countries, since the entire planet is their mother country and only the wise men direct and advise them.

I asked Salvador about religious matters, and the answer was that on Venus religions do not exist, since each citizen behaves on the street as if they were within a temple. Each person on Venus considers that the temple exists within us.

The sidewalks of the streets in the cities of Venus are not still, because they are made of metallic bands that are in constant movement, thus saving the effort of the pedestrians. The flow of their streets—that is to say, the center of their avenues—has metallic belts that collect the force of the sun with which their cars are propelled, thus people never walk on the central flows of their street.

So, on Venus everything belongs to everybody, and the entire Venusian family is a great family. Children are born in special maternity rooms and they are educated and raised in collective homes. When a child is born, they stamp a mark on his foot; that mark indicates the child's origin and faculties. Thus according to this mark the child is educated in the collective home. When the person reaches maturity, he passes to occupy the place in society that corresponds to him. So, on Venus children are not wandering on the streets; the government manages them until they reach a suitable age to be classified according to their physical and mental qualities, and assigned to whatever place they are needed.

In these conditions, private families do not exist; everybody in Venus belongs to a single united family. There is no hunger, or war, or social classes. There only reigns wisdom and love.

For five days he lived on the planet Venus, and returned to the Earth after having verified the reality of all the affirmations made by the Venusians. The Venusian civilization is a million times more advanced than ours, proud Earthlings. Salvador narrated what he saw; we limit ourselves to comment about it.

On that planet he found two French residents, twins and veteran brothers of World War II. They also were transported to Venus, and they begged and cried to the Venusians to not take them back to the Earth; now they live happily.

The Venusians affirmed that some of them are here on our planet Earth dressed as any citizen in order to study the humanity of our planet. They stated that many thousands of years ago they lived the historical stage that we Earthlings are passing through now. They also knew the wars and the astute political leaders, until brotherhood was born. Nowadays, they do not have flags. They have made of their world a single mother country and are governed by wise people who limit themselves to advise them with wisdom and love.

Salvador came back to the planet Earth; he was brought back in order for him to inform the inhabitants of the Earth that Venus is inhabited.

Philips Laboratories analyzed the soil and plants in the place where Salvador saw the spaceship, and they found a strange molecular and atomic disorder.

The North American scientist George Adamski—who met some Venusians in the desert of Nevada—came in contact with Salvador and gave a lecture about this subject in the theater Insurgentes of Mexico City.

Great German scientists investigated the land where Salvador found the spaceship and the outcome of their investigations was the same as Philips. Another great scientist came from the palace of the kings of England in order to investigate the case, and the conclusions were the same as Philips.

So the German commission of scientists—interested in this matter—visited Salvador and studied the region of the events. Thus, there are no doubts about it; however, the imbeciles will continue to laugh as always, because they are imbeciles.

In these difficult times in which we live, we will be helped by the inhabitants of other planets. It is necessary for us to learn how to telepathically communicate with them. Jesus said, "Ask, and it shall be given you, knock and it shall be opened unto you" [Matthew 7:7]. All of us can visit other planets if we know how to ask.

Gnostics must develop telepathy. Gnostics must go to the fields, to the deepest forests, and peacefully and in deep meditation, telepathically enter into communication with Venusians, Mercurians, or Martians and request to be taken to Venus, Mars, or Mercury.

In the peace of the mountains, or in a solitary beach, any day we can have the happiness that Salvador Villanueva Medina had. Each of us can be taken to Venus or to other planets. The system to communicate with those Human Angels is telepathy. Our Universal Christian Gnostic Movements have formidable systems in order to develop telepathy.

Whosoever wants to visit other planets must not drink alcohol, smoke, or have any vice. Our international Gnostic missionary J. A. B. was visited by an interplanetary spaceship during a retreat in the Summum Supremum Sanctuarium.

Yes, the thought waves of the suppli-
cant travel to the planet Venus in a few
seconds, and if we are worthy and deserv-
ing, we can receive an answer.

Thus, any given day, in the solitude of
the fields, we can have the happiness of
seeing a spaceship landing near us and
then be taken by Mercurians, Martians,
etc. since they are true humans with bodies of flesh and bones, humans with the
souls of Angels: Human-Angels.

Chapter 9
The Pluralized "I"

There is energy that is free in its movement, and there is static energy. The "I" is a knot that must be untied. The "I" is static energy. The universal Spirit of life is energy that is free in its movement. The spirit is not the "I." The Soul is not the "I." The physical body is not the "I."

It is necessary to know that the "I" is the Satan of which the Bible speaks. The "I" is Ahriman of the Persians. The "I" is a handful of memories, desires, passions, hunger, fears, etc. A so-called "superior I" does not exist. Indeed, our real Being is beyond any type of "I." Our real Being is the Being and nothing but that: the Being.

The soul is the Being, the Spirit is the Being, but the "I" is not the Soul or the Spirit. The "I" is the devil, and that is all. The "I" exists in a pluralized manner: with this we mean that the "I" is a legion of devils.

Just as water is made up of many drops, just as the flame has many sparks, likewise the "I" is made up of many small "I"s.'

Each desire is personified by a small "I." Each appetite is personified by another small "I." The seven capital sins are personified by seven "I's"—one for each capital sin, seven for the seven capital sins.

All vices, passions, and evilness are personified by small "I"s that in their conjunction constitute the "I" or reincorporating ego. What reincorporates is the "I." The "I" reincorporates in order to satisfy desires and to pay karma. The "I" is the origin of pain; the "I" is the origin of all of our evilness.

When the "I" is reduced to dust, what remains within us is the Soul. Indeed, the nature of the soul is happiness. The soul is happiness. To seek outside for happiness is absolutely absurd, since happiness comes to us when the "I" has died. Therefore, while the pluralized "I" continues to exist, there can be no happiness.

In life, there are pleasant hours and enjoyments, nevertheless, happiness cannot exist as long as the "I" is not dissolved.

When the "I" is reduced to dust, we can then reincarnate on other more advanced planets in order to work on the realization our inner Self.

The dissolution of the "I" brings true freedom.

The Venusians are truly happy because already annihilated the "I"; they do not have "I."

The Venusians do not need money because they do not have anxieties for accumulation. Since they do not desire anything, they do not have greed; they are satisfied with their daily bread. That kind of consciousness is possessed by Beings who no longer have the "I."

Authorities are not necessary in Venus because there is no violence. Only the "I" is violent. Government is not necessary in Venus because each citizen knows how to be governed by his Self. When the "I" has been annihilated, each citizen becomes governed by his Self; then, who wants to govern others?

On Venus, private families do not exist. All the Venusians form a single family; this is only possible thanks to the fact that they already annihilated the horrible pluralized "I." The "I" is what we call "my" family, "my" house, "my" properties, "my" lust, "my" resentments, "my" desires, "my" passions, "my" memories, etc.

The "I" continues in our descendants. The "I" is the race, the nation, my social class, my money, my family, my inheritance, etc.

The "I" is the subconsciousness. When the "I" is annihilated, the subconsciousness becomes consciousness.

We need to annihilate the "I" in order to transform the subconsciousness into consciousness. Only by annihilating the "I" can the subconsciousness become consciousness. When the subconsciousness becomes consciousness, the problem of astral projection has been resolved.

When the subconsciousness becomes consciousness, we no longer needed to worry about astral projection, because while the physical body sleeps, we live in an absolutely conscious manner within the internal worlds.

At the present time, humanity is ninety-seven percent subconscious and only three percent conscious. We need to become conscious one hundred percent.

The inhabitants of Venus are one hundred percent, totally conscious. The humanity of Venus annihilated the "I."

Indeed, the "I" can only be annihilated by means of a rigorous creative comprehension. We need to perform a self-dissection of the "I" with the scalpel of the self-criticism. Instead of criticizing others, we must criticize ourselves.

Practical life is a full-length mirror where we can see ourselves as we really are. When the mind is in a state of alert perception while coexisting with the neighbor, we can then easily discover our defects because they spontaneously arise. Our defects emerge in an astounding manner while in relation with the neighbors, with our friends, with our fellow workers, with our spouse, with our children, with our acquaintances, etc., thus, if we are alert and vigilant as watchmen in the time of war, it is clearly logical that we can see our defects as they really are.

Self-discovery exists while coexisting with others, if we are in the state of alert perception.

Any discovered defect must be intellectually analyzed; nonetheless, the intellect is not everything. The intellect is only a fraction of the mind.

We need to delve; we need to explore the subconsciousness more deeply in order to discover the inner roots of our defects. Yes, only through very deep meditation can we really explore the subconsciousness. Thus, when we have comprehended a defect completely, the energetic "I" that personifies it can be disintegrated. Thus, this is how we die from moment to moment.

We need mystical death. We need the death of the "I." Let us remember that each one of us carries a legion of devils within. Yes, the "I" is a legion of devils. The body of desires exists within each person and within the body of desires exists the pluralized "I."

The pluralized "I" miserably squanders the Essence, that is to say, the raw matter, the substance of the soul. The "I" squanders the precious psychic Essence in atomic explosions of anger, greed, lust, pride, laziness, gluttony, etc.

When the "I" is dead, the Essence is accumulated and becomes the soul. We need the death of the "I"; we need only that which is happiness to live within us, that which we call soul. When the "I" dies, karma ends, and in fact we become free.

The inner contradictions of each person are due to the pluralized "I." I.e. "I am going to read the newspaper," says an "I" in the intellectual center. "I do not want to read, what I want is to ride a bicycle," says another "I" in the motion center. "I love that woman, I adore her," says an "I" in the center of emotions. "I do not love her, what I want is money," says another "I" in the mental center. "To the devil with these preoccupations, what I want is to eat," says the "I" of gluttony through the digestion system, "and I want to eat a lot," says the "I" of greed... "I swear to be faithful to Gnosis," says the mystical "I." "To the devil with Gnosis," furiously exclaims the intellectual "I." "It is better to get money," says the "I" of greediness. And, "I will affiliate myself to another school, better than Gnosis," says the "I" of curiosity.

Thus, this is how we do not have individuality; yes, we are not individualized because we are legion of devils. Thus, when the "I" is dissolved, individuality—

that is to say, the individual soul—is the only thing that remains within us.

The Venusians are true, sacred individuals, because they do not have "I." The Venusians are truly perfect human beings. We Earthlings are intellectual animals, because we do not have authentic individuality.

We have seen too many people swearing fidelity to Gnosis, giving an oath before the altar, and thereafter in time they go to another school and declare themselves enemies of Gnosis. This is because they do not have individuality. The "I" that in a given moment was enthusiastic about Gnosis is later displaced by another "I" that detests Gnosis.

Present "human beings" cannot have continuity in their purposes because they do not have individuality. They are legions of devils, and each devil has its own criteria, ideas, opinions, etc., etc. The present human being is a non-realized being.

We still do not possess the Being, and only the Being can grant us true individuality.

Chapter 10
A Jupiterian Visitor

The news that arrives from all over the world assures that the cosmic ships land in different places of the Earth.

What bothers the intellectual-loafers the most is to not be able to capture one of those ships, with crew and everything. We are absolutely sure that the cannibals from Africa and the Amazon also felt very annoyed when they could not manage to capture an explorer.

Regarding concrete facts about flying saucers, people want to proceed like cannibals, however, it is clear that the crews of those cosmic ships—who know human savagery very well—are not willing to let themselves be ensnared, since they know very well the fortune that awaits them, that is, the intellectual-loafers would make them prisoners, the ships would be seized and used for war, etc.

Yes, the crews of those cosmic ships are not willing to serve as "guinea pigs," therefore, before letting themselves to be trapped—similar to those explorers from the white race that flew past the tribes of cannibals—they prefer with good reason to disappear within the infinite space.

We are stating something that may hurt the intellectual-loafers very much, since they love themselves too much, boasting about themselves as super-civilized, even if in their depths they are truly nothing but savages dressed in the modern style.

In Brazil, near Parana, a cosmic ship landed before the presence of a famous scientist, whose last name is Kraspedón. The captain of that ship invited the mentioned scientist to visit his ship.

The mentioned scientist not only came to know the interior of the ship, but also its crew.

The captain of the ship said that he came from a satellite (a moon) of the planet Jupiter. He spoke perfect Spanish, and promised the mentioned scientist to visit his home in return. When Mr. Kraspedón wanted to give his home address to the captain, he replied that it was not necessary; when refusing, the captain said, "We know perfectly well how to find you on the Earth."

Six months later, on a Sunday, Mr.
Kraspedón—withdrawn in the study room
of his house—was suddenly interrupted
by his wife, who informed him that at the
door of his house was a man who wished
to speak with him. She said that the man
was carrying a Bible between his hands
and insisted on giving explanations about
it. Mr. Kraspedón commanded his wife to
dismiss the visitor and to close the door.
Moments later, the lady returned and
informed her husband that the visitor
did not want to go away and insisted on
speaking to him.

A little sour-faced, the scientist resolved
to leave his study and go to the door in
order to take care of the visitor. Great was
his surprise when he found himself face
to face with the captain of the cosmic
ship whom he had met six months earlier.

Mr. Kraspedón welcomed the visitor,
and invited him to come inside the living
room of his house. Soon their conversation began. The mentioned scientist
wanted to examine the intellectual capacities of the Jupiterian, placing him, so to
say, in a very difficult position, by means
of complicated questions about the Bible.

The visitor demonstrated his possession of a most illuminated intelligence, since he knew even the deepest roots of Greek, Hebrew, and Aramaic, thus, he knew how to give the sacred scriptures highly scientific, deeply philosophical, extraordinarily artistic, and transcendentally mystical interpretations.

After the interview, there were two more interviews in different places of the city to which the mentioned scientist attended, accompanied by a professor of physics and mathematics.

The teachings that the Jupiterian gave on the subject of astronomy were indeed formidable; all of that knowledge is transcendental.

Mr. Kraspedón is a serious scientist; he is not a charlatan. Thus, he resolved to condense all of the knowledge that the Jupiterian gave him in a precious book written in Portuguese and entitled "Flying Discs."

The Jupiterian warned that the atomic explosions are altering the superior layer of the terrestrial atmosphere. This layer is the supreme filter that alters and disarranges the solar rays, transforming them into light and heat.

The Jupiterian said that if atomic scientists continue with their nuclear blasts, the day would arrive in which the supreme filter would be incapable of disarranging and altering the solar rays into light and heat; then we would see *the Sun black as a sackcloth of hair, the Moon red as blood* [Revelation 6:12], and a ferruginous red color upon the face of the Earth.

The Jupiterian warned that when the superior layer of the terrestrial atmosphere—that serves as support for the life of the Earth—becomes disturbed, great earthquakes would occur and the cities would fall to the ground, like castles made of cards, turned into dust.

The Jupiterian gave an alert that the navigators of outer space, who often visit the Earth, are already seeing the superior layer of the terrestrial atmosphere with a clearly altered process, and without the brightness and splendor of other times.

The Jupiterian said that atomic radiation from thermonuclear war will saturate the water that we drink, the harvests

Between 1945 and 1998, there were approximately 2,000 *reported* nuclear explosions. This number does not include classified military information.

whereupon we maintain ourselves, the clouds that bring us the rains, etc.

The Jupiterian warned that atomic radiation will damage the phosphorus in the brain of human beings, and everywhere we will seen Dantesque scenes in the streets, hospitals full of people, multiplication of cancer and leukemia, millions and millions of dead, hunger, and desperation.

Time is passing; now, the atomic blasts proceed underground, as much in Russia as in the United States. France and China continue making atomic explosions in the atmosphere, and newspapers from the entire world bring news of sudden earthquakes, either in Chile or in El Salvador, now in Iraq, Japan, etc.

We are facing concrete facts that are impossible to deny.

Envious people will be bothered very much by the narration about the Jupiterian and Mr. Kraspedón, thus, it would not be very strange to us for them to send against us all their satires based on their stupid skepticism, similar to the satires of those who mocked Pasteur, Galileo, Edison, etc.

What really bothers the envious people is to not have the opportunity that Mr. Kraspedón had. We are sure that if the same opportunity were granted unto these envious people, they would abuse it, inevitably capturing or killing the visitors from the infinite cosmos.

Nevertheless, cannibals are cannibals, and the inhabitants of other planets know very well how to protect themselves from them, thus disappearing into outer space before the barbarian hordes of Earthlings can capture them.

Chapter 11

The Gnostic Movement (1964)

Earthlings have directed themselves towards the conquest of outer space, and they do not care at all about flying saucers or spiritual matters.

If we compare the flying saucers with the space rockets, the rockets are grotesque and ridiculous. Nonetheless, the Russians and the "gringos" want to conquer the Moon, and it is clear that they will land on the Moon with them.

What is most lamentable of all this is the aggressive instinct of the current terrestrial humanity. Whoever dominates the Moon will want to turn it into an armed military platform with atomic potential.

They have not landed on the Moon [as of 1964], yet, like the Tyrians and Trojans [enemies in *The Aeneid*] they are already talking about orbital nuclear missiles in order to destroy defenseless cities. Yes, this is the regrettable state of the current terrestrial humanity.

The inventor of the hydrogen bomb still does not know the damage that he caused to humanity. If one of those hydrogen bombs exploded in the superior zones of the atmosphere where there is a deposit of pure hydrogen, then the entire atmosphere of the Earth could burn, fulfilling the prophecy uttered by the Apostle Peter in his second Epistle that states the following:

> *But the day of the Lord will come as a thief in the night; in the which the heavens shall pass away with a great noise, and the elements shall melt with fervent heat, the earth also and the works that are therein shall be burned up.* - 2 Peter 3:10

Photo of Starfish Prime, a hydrogen bomb exploded 250 miles above the Pacific Ocean by the United States in 1962. This was part of a series of tests conducted during the Cold War by the United States and the Soviet Union. Image from U.S. Congress "Report of the Commission to Assess the Threat to the United States from Electromagnetic Pulse (EMP) Attack, Volume 1: Executive Report, 2004"

It is clear that before this could happen, before any wacky Earthling could conceive the sick idea of making this detrimental experiment with the hydrogen bomb, before it could explode in the superior zones of the atmosphere amidst the pure terrestrial hydrogen, the deposit of living universal hydrogen, we can be sure that they, the navigators of infinite space, would make the planet Earth explode into pieces, so the humanities of other planets of the solar system will not have to undergo the consequences of the explosion of our terrestrial hydrogen. Yes, such a terrestrial catastrophe would frightfully affect the other planets of the solar system. Thus, before this could happen, they, the navigators of the sidereal space, would be forced with deep pain to destroy this planet, making it to fly into pieces, because it is not right that other planetary humanities should suffer the consequences of the madness of the Earthlings.

At these moments, the inhabitants of the Earth are full of pride and arrogance. The perverse Earthlings have raised the "Tower of Babel" with which they think to conquer outer space, thus the inhabitants

A promotional image of the intention to colonize Mars.
IMAGE CREDIT: NASA

of the other planets of the solar system already have orders to defend themselves.

Space rockets will inevitably land on the Moon. Later on, these Earthlings perversely and arrogantly will launch themselves to Mars.

By logical deduction, the encounter with humanities of other planets is totally inevitable, thus this failed and degenerated root race has no alternative but to transform themselves or perish.

The Universal Christian Gnostic Movements want to forge groups of men and women of good will in order to welcome our brothers and sisters from space.

Millions of Earthlings—filled with pride, arrogance and perversity—believe only in their space rockets and their weapons of mass destruction. Those hordes of learned ignoramuses are surrendering themselves to all the vices the Earth, and they laugh at everything that smells of flying saucers.

We, the Gnostic people, do not in any way accept the Antichrist of false science, neither we can even slightly think that the infinite space can be conquered without having previously conquered ourselves.

It is completely absurd to suppose that the barbarian hordes of Earthlings can conquer other planets and enslave superior planetary humanities.

The Gnostic people do not accept the perversity of the intellectual-loafers, thus they have resolved to organize an army for world salvation with men and women of good will who are willing to joyfully welcome our brothers and sisters from sidereal space.

The Universal Gnostic Movements will be established everywhere, in different places of the Earth, true mystical cenacles conducive to the study of the cosmic laws, in order to prepare people so that they can welcome our brothers and sisters from space.

Yes, the Gnostic Movements comprehend the necessity of cosmic ships in order to travel throughout the infinite; nevertheless, we do not believe that the way of perversity, pride, and rockets is the correct way.

The infinite cosmos is sacred, and is governed by divine laws that cannot be broken with impunity without receiving the disastrous consequences. We, the

Gnostics, are willing to study the laws of eternal space with all humility and at the feet of our brothers and sisters from space. We know that this is the precise way that can allow us to navigate in true cosmic ships through infinite space.

Now we need to be prepared, by finishing with all our defects, by dissolving the "I" that we carry within, that detrimental Mephistopheles.

In some places of the Earth, very secretly, there are select human groups in contact with the cosmic visitors, from whom they have received a small number of interplanetary ships. In the Himalayas, filled with snow and cold, there is a certain group of Lamas who are in contact with our brothers and sisters from space. That group very secretly possesses a certain number of flying saucers with which they travel through the infinite. In other places of the Earth there are similar groups that already possess flying saucers. We are affirming something that cannot even remotely be accepted by the know-it-all, ironic, sarcastic, and vain intellectual-loafers. However, what does it matter to science or to us?

Let us go, brothers and sisters of the Gnostic Movement, to humbly prepare ourselves in order to become worthy, and as a select and hidden group to meritoriously receive our brothers and sisters from space.

On December 27, 1968, our Second Gnostic Congress will be celebrated in Barranquilla, Colombia, South America. This event will be international, thus all the delegations from all the Gnostic sanctuaries of America must attend. Among other things, this Gnostic Congress must deeply study methods and systems of conducive action to propagate these ideas everywhere, in order to prepare the environment for our brothers and sisters from space. It is necessary, it is urgent, to propagate these teachings and to form true homes for our brothers and sisters from space. We invite the brethren of all religions, schools, and sects to this great international, spiritual, ecumenical congress, without distinctions of creed, race, society, school, order, sexes, or color.

Chapter 12

The Extraterrestrials' Mission

Friends, I address all of you again in order to converse a little about the cosmic ships that navigate outer space. This very intriguing theme about flying discs has been propagated throughout the Earth; already, nobody can deny it. Unquestionably, today the one who dares to deny it is heard only because of his stubbornness.

The British themselves no longer deny it. In England a short time ago, they officially declared, "The flying saucers exist and are manned by extraterrestrials, who are people who supersede us by many millions of years of civilization. Since we, Earthlings, cannot comprehend them, we prefer not to think about it. We are going to see if we will be capable of making our own ships in order to conquer the infinite." These are the words that the British gave to the world regarding flying saucers. Therefore, those who today deny it are presumed to be stubborn as described, since this has been demonstrated over and over again and with photographs, as much in the north as in the south, in the east as in the west of the world.

Obviously the Earth cannot be the only inhabited planet. It would be an absurdity to think that our world, a very small speck of sand within the infinite space, would be the only planet with the exclusivity of having people.

Indeed, truly, the plurality of inhabited worlds, maintained some time ago by Camille Flammarion, is a tremendous reality. Nevertheless, scientists, as usual, continue doubting.

Recently, a space probe was sent to Mars in order to know if there is life there. The scientists of NASA concluded emphatically affirming that there is no life on Mars. Yet, the photographs that they showed humanity are not of Mars; indeed, those photographs are of the Moon. This means that the Martians knew how to adjust the American photographic apparatuses towards our terrestrial satellite. Therefore, what those machines transmitted to the Earth were lunar images. This, that I know, is simply

An image from the NASA Phoenix mission.
Image credit: NASA/JPL-Caltech/University Arizona/Texas A&M University

based—even if it seems incredible—on the information given to me by an extraterrestrial who laughs when commenting about the foolishness of the Americans and their NASA.

It would be an absurdity to suppose that people as cultured as the Martians would let Earthlings develop a detailed map of their planet; they know very well the intentions of the Earthlings. They do not ignore the destructive character of the inhabitants of the Earth. Yes, Earthlings are destructive; they have demonstrated it to satiation. Thus, nobody in the cosmos ignores this.

It is not irrelevant to remember the atrocities that the terrible Spaniard Hernán Cortes committed here in our beloved Mexico, or those that Pizarro committed in Peru. Therefore, if the Martians were invaded by Earthlings, such would be the fate they would experience, and thereafter their wonderful ships with which they navigate the infinite space would be used with Machiavellian intentions by the governments of Russia and the United States.

Such ships would be loaded with atomic bombs in order to destroy defenseless cities; they would be used in order to conquer other planets of the starry space, thus exporting all of the Earthlings atrocities across the cosmos; this is very well known by the Martians, thus, they are not so ingenuous as to let us make a map of their planet.

What I am affirming is based on trustworthy information; I am not trying to invent anything new. The Martians have orders to defend themselves, and they will do so if the Earthlings try to invade them.

On Mars there are cities like Tanio, for example, which is inhabited by pacific people who never prepare plans of war nor do they invent atomic bombs in order to destroy anybody. The inhabitants of Tanio are in no way willing to let themselves be invaded by hordes of Earthlings. This is the crude reality of the facts.

Multiple ships navigate the starry space. There are gigantic ones like the cargo ships that carry small ships within their belly. These cargo ships are used for trips not only within our galaxy, but also to navigate to different galaxies; they are properly prepared to travel throughout the unalterable infinite.

There also exist tiny ships—and this is something that will surprise you—of about twenty or thirty centimeters. Absurd, you might say, such tiny ships do not exist! Who could navigate in such small ships? They are navigated by cosmic lilliputians, Supermen with very tiny gelatinous bodies, humans who are no more than ten centimeters in size.

To what will one be exposed when affirming such a thing in this day and age, in this twentieth century, in the atomic age, an age of X-rays and laser rays? One is exposed to mockery, of course, since the scientists of our planet Earth believe that they own the entire wisdom of the universe, but they are deluded.

All of the rottenness of theories that abounds here, there, and everywhere and that forms the culture of this century is indeed vulnerable. That is not pure science. The scientists of the Earth do not know pure science, since in order to have access to the amphitheater of cosmic science, pure science, it is necessary to have opened the Inner Mind.

Allow me to remind you again that there are three minds in the human being. The first one is the Sensual Mind, where the leaven of the materialistic Sadducees is deposited. This type of mind develops its basic concepts via external sensory perceptions; this mind knows nothing about reality, nothing about that which is beyond simple external perceptions.

The second mind is the Intermediate Mind, where all types of beliefs are deposited. Obviously, to *believe* is not to *know*. We have entered into the age of knowledge, in the Age of Aquarius; beliefs are beliefs, yet they do not imply wisdom. The leaven of the Pharisees is deposited in the second mind; Jesus the Christ warns us against the leaven of the materialistic Sadducees and the leaven of the believer Pharisees.

If in fact we want to penetrate in the amphitheater of pure science, a third mind is necessary. This third mind exists but it is closed. We need to open it. This is the Inner Mind. This mind is opened when one finishes with one's psychological defects: when anger, greed, the frightful lust, envy, pride, laziness, gluttony, vanity, etc. are eliminated in oneself.

Indeed, truthfully, those who eliminate their psychological defects will awaken their consciousness. The awakening of the consciousness opens the Inner Mind. When the Inner Mind is opened, true faith arises, which is not the run-of-the-mill faith, but the conscious faith of the one who knows, of the one who can see, hear, touch, and feel the great realities of the amphitheater of cosmic science.

The extraterrestrials are people who have opened their Inner Mind: Supermen in the most complete sense of the word, as those of "El Desierto de los Leones," a wonderful contact that I recently had and that I narrated to you, because for the good of humanity one must give testimony of what one has seen and experienced. So, I will never be ashamed of having given such a testimony before the solemn verdict of the public consciousness.

Regarding the promise of being taken to other planets that the captain of that intergalactic ship gave me, "Along the way we will see," that promise was convincing for me, because I know that they never deceive anybody and they always fulfill their word, no matter what the cost might be. They speak little but they say much, and when they utter it they fulfill it, since they are not Earthlings.

So, I comprehended that such a way is the path of wisdom, the path of perfection. I will fight, yes, in order to eliminate my psychological defects.

These extraterrestrials, I understood, are intergalactic travelers. The small ship that I saw and within which they descended, obviously returned to a cargo ship that had been in orbit around the Earth. Supermen navigate in those cargo ships, from galaxy to galaxy; they are infinitely perfect and are beyond good and evil.

Earthlings abhor them. A short time ago, when two of those ships soared over the United States, jet fighters left in order to destroy them. One of the ships flew away and vanished into space; the other descended, settled smoothly on a tower of electrical energy, then the blackout of New York took place and not only affected the American nation but also Canada.

Generals of the American army exclaimed, "Behold, the Achilles' heel of the United States of North America." Indeed, the powerful titans of the north can do nothing without electrical force, thus just a handful of extraterrestrials manning a ship paralyzed the powerful nation of the American states.

Therefore, if the extraterrestrials had wanted to destroy that country they would have done it in a matter of seconds. Likewise, if they wanted to make the planet Earth explode asunder into pieces, they would already have done it. But they are not destructive. They love us and come to us in order to help us, yet instead of welcoming them with our open arms like our saviors, like our redeemers, instead we receive them with cannon shots, or we flee desperately, as the cannibals in the mountains flee when they see an airplane. Yes, this is the lamentable state in which we are. The extraterrestrials are cultured people who do not kill anybody, not even a small bird, and Earthlings fear them.

Many ask, "Why don't the extraterrestrials land in cities like New York or Paris and show themselves in public in order to dictate lectures?" To such a question I would respond the following: If somebody found himself in a deep jungle surrounded by cannibals, what would he do? Indubitably, he would flee, he would have no other alternative.

The extraterrestrials could defend themselves, who can deny it? Nevertheless, they do not have the desire to destroy anybody; they are not assassins. Earthlings are mistaken when thinking that they come to assassinate people; such a thought is not true.

We do not deny that on some occasions they have taken some people into their ships, then to space, and thereafter they have brought them back, returning them to the place where they took them; yes, that is true. However, this

has an explanation: it so happens that
Earthlings are very strange; they have
their consciousness asleep. They seem
like somnambulists who walk the streets
with their tremendous perversity; indeed,
Earthlings are somehow a source of curiosity, therefore some are taken in order
to be studied in laboratories that exist
in space. These laboratories are located
within some cosmic ships. So, beings as
rare as sleepy, unconscious, and destructive as Earthlings are a cause for curiosity,
and for such a reason Earthlings are taken
in order to be placed in laboratories for
study. That is the crude reality. However,
no type of harm is ever inflicted upon
them, and they are brought back to the
place from where they have been taken.
It is clear that there are some exceptions,
about which I will speak to you tonight.

An unusual but wonderful case happened in Ecuador. One day among many,
a man who had studied in some Eastern
schools saw a cosmic ship that hovered
over the garden of his house. Certain
extraterrestrials opened the hatchway of
their ship, descended by a staircase, and
approached him; they invited him to
board the ship, and he accepted. He was

a cultured and spiritual man in the most complete sense of the word. He was willing to go, so that when he was invited to take a stroll in space, he accepted.

He was taken to a satellite of Jupiter, to Ganymede, where he witnessed a powerful civilization.

The inhabitants of Ganymede have a brain that is little more voluminous than those of the Earth. Their pineal gland is connected to their pituitary gland by means of certain nervous threads, and their pituitary gland in turn is connected to their optical nerve by other nervous threads. Consequently, the inhabitants of that satellite have a sixth sense with which they can see the fourth dimension and even more: the fifth, sixth, and seventh dimensions of nature and the cosmos.

They build their houses underground, they have rich agriculture, and they do not have animals because the atmosphere there is not favorable for the inferior species. Water is removed from certain volcanoes and with it they satisfy their needs.

All the inhabitants work in their factories. There is no money, it is unknown there. Thus, in exchange for work, all

Ganymede is the largest moon in our solar system, and is bigger than Mercury and Pluto. Scientists have found evidence of an atmosphere and an underground ocean.

the inhabitants have bread, shelter, and refuge. The cosmic ships are the property of all. They lack nothing. No, they do not need that element called money that has inflicted so much damage upon the inhabitants of the Earth. Thus this is how the inhabitants of Ganymede are.

Since they possess a sixth sense, they have studied medicine in a perfect manner, not only in its physical-chemical and biological aspects, but also in its psychic and vital aspects. They know interior and exterior anatomy, which unfortunately the scientists of Earth do not know.

The hour has arrived to understand that Ganymede is a satellite that rotates around the titan of the heavens called Jupiter. Jupiter has twelve satellites; it is as if it has a new Solar System for itself.

Often I have personally observed Jupiter through a telescope; I have seen its two belts at the equatorial center of that wonderful world; those belts with their satellites rotate around that planet. Jupiter is extraordinary, in the most complete sense of the word, a jewel of the heavens.

Previously, the inhabitants of Ganymede lived in a world that was called the "Yellow Planet." It is good to know that in our Solar System a planet once existed on which its people devoted themselves to atomic experiments. They performed multiple experiments; they made bombs that were successively more destructive, and they finally manage to burst that planet asunder into pieces. Some fragments still rotate around our solar system—asteroids, we would say, or loose earth—and this is known by the astronomers.

But before that catastrophe happened to the Yellow Planet, the inhabitants of that world had been facing the philosophical dilemma of "to be or not to be." A great Avatar or Messenger warned them about the catastrophe that waited for them. Most of the people, as always, did not believe, yet others accepted it; then they were affiliated to the teachings of that wise man. They were sufficiently prepared by him, and finally they were psychologically corrected, and ready to investigate Ganymede. Before the catastrophe happened, that wise man took them to Ganymede and they settled there.

Well, the friend of whom I am speaking was taken to Ganymede, and was some days there. His diseases disappeared; he revitalized himself. He was placed under special scientific treatments. They invited him to live among them, and he accepted on the condition that they allowed him to return to the planet Earth in order to give his goods to his brother. He returned and gave his entire fortune to his brother and his brother's wife. He determined a certain date to say farewell to them, since he was going to make a long trip. On that date, where he was preparing himself to leave, a gorgeous ship landed, which illuminated the center of the garden.

"I am leaving," he said to them. His brother and sister-in-law were astonished.

"Ah! I had a feeling about this," his brother said.

He boarded the ship, and at that moment exclaimed, "I am leaving by my own will, far away from this planet Earth!" Thus, the ship took off.

Nevertheless, a small camera in the form of television was left for his terrestrial brother on Earth. It was enough to move some buttons and it worked. The solar energy animates that apparatus; it is enough to concentrate on certain antennas that it has, in order to establish contact with Ganymede. They communicated with the inhabitants of Ganymede and their brother, hoping to be taken away someday; thus, finally such a longed for day arrived and they were taken away.

The inhabitants of Ganymede have an extraordinary wisdom, and they are willing to help the Earthlings; they know the state in which Earthlings are. They do not ignore that a great catastrophe approaches them, that a giant of the heavens called

"Hercolubus" is traveling at vertiginous speeds through the starry space.

When Hercolubus appears and all of you can see it in plain sight, you will then be convinced about what I am stating to you. Then, that planetary mass—which is enormous—will magnetically attract the fire of the interior of the Earth, which will bring forth volcanoes everywhere, and terrible tsunamis and great earthquakes will be originated. The entire crust of the Earth will be destroyed, burned, and incinerated, and a revolution of the axes of the world will be produced at the height of the approach of Hercolubus; the poles will become equator, and the equator will become poles. Then, the waters of the oceans will sink these continents.

This is how this perverse humanity will end, a humanity that gave themselves to all the vices, that degenerated with homosexuality and lesbianism, a destructive humanity, where each country rose against the other, where each human being raised his hand against his brother. This is how this humanity will soon end.

There will be survivors, yes, there will be. This is very well known by the inhabitants of Ganymede. They will continue—little by little—taking the more select people to their world. There, they will cross them with people who live on their planet, and from such crossings a very special type of humanity will be born that will be returned to Earth. Wonders will be made with that type of humanity, because they will develop the faculties that the inhabitants of Ganymede possess.

A lovely opportunity will be given to some survivors who will remain on our planet Earth on an island of the Pacific, and there the sixth great root race will be born. I say sixth because there have been five root races on the face of the Earth, and all of them have ended with great cataclysms.

Let us remember the Atlanteans; they perished within the waters of that great universal deluge produced by the revolution of the axes of the Earth, where the seas swallowed Atlantis with its millions of inhabitants. That was, I repeat, the universal flood. Thus, this is how the fourth great root race perished.

We are the fifth root race. It is obvious that the people of this root race will perish by strong earthquakes and great tsunamis. This is why it is written in the Aztec Calendar: "The children of the Fifth Sun will perish by the fire and earthquakes."

Friends, at this moment in which we talk about very interesting aspects related with the extraterrestrials, another case comes to my memory about a man named Mr. Salvador Villanueva Medina, who went to Venus. I know him; he is my personal friend.

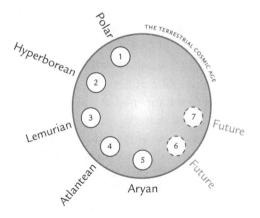

This planet has been home to more than this current humanity, who is the fifth major race of human beings of the current cosmic age. See the Glossary entry for Aryan. Read *Gnostic Anthropology* to learn more.

This friend of mine was driving a car towards the United States of North America. The car was damaged. He went underneath the car in order to discover what happened to it. Then he heard steps on the sand... when coming out from beneath the car he saw two men of average stature. They invited him to follow them. He did, and when arriving at a place where a cosmic ship was standing, the two enigmatic personages invited him to come inside the ship. Thus, once inside the ship, the hatchway closed and soon they departed to the planet Venus. He was there for five days, he saw that powerful civilization, and thereafter they brought him back to the planet Earth.

In the republic of Ecuador, another unusual case happened. An incredulous, skeptical, and materialistic man was seated at the end of his garden. A ship landed a short distance away; they invited him to board the ship, and he was taken to the planet Jupiter. There he remained for several days. He witnessed an extraordinary civilization. They invited him to stay, yet he said to them, "I must return to speak with the Earthlings, and give testimony about what I have seen and heard here." He returned to Ecuador and since then, that man dedicated himself to study Gnosis.

The extraterrestrials exist and will always exist. Before the inhabitants of the Earth learned the first foundations of mathematics, the space travelers had already come through the unalterable infinite. Thus, if they wanted, they could have destroyed us and made the Earth explode asunder into pieces, but they are not destructive. They come with the intention of helping us on the eve of the great cataclysm that is approaching. They come to help us; some of them have remained hidden in diverse places of the Earth, and only wait for the opportune moment in order to enter into activity.

For example: in the Antarctic in an underground city lives a group of extraterrestrials. They are humans who came from the Blue Galaxy, humans of blue skin. Obviously, they have built that city under the ice; they have all manners of comfort, an atomic lighting system, etc. They enjoy beauty, are wise by nature, and a very wise king governs them.

Soon they will come; when the catastrophe approaches, these humans of the Blue Galaxy will walk on the streets of the cities trying to aid those who remain. Before the great universal fire devours the face of the Earth, these humans of the Blue Galaxy will appear in order to teach the path of uprightness to this humanity. Those who listen to them in those days will be totally saved.

Chapter 13

Answers about Extraterrestrials

Audience: An Earthling had the opportunity to see an extraterrestrial. Thus, he asked, "How many planets does our solar system have?" The extraterrestrial answered that there are twelve planets. I want to know if this is true or not.

Samael Aun Weor: Well, indeed, this time they did not gave you the exact number of planets, because there are thirteen planets in our solar system. Let us count them in sequence so that this will be understood: Earth, Mercury, Venus, Sun, Mars, Jupiter, Saturn, Uranus, Neptune, Pluto, Vulcan, Persephone, and Clarion: a total of thirteen worlds in our solar system! There were fourteen, but it so happened that the inhabitants of that Yellow Planet or planet fourteen dedicated themselves, as I already said, to atomic experiments, thus they totally destroyed their world. Now, loose earth or asteroids rotate around the sun, and are remnants of that destroyed planet; this is known by the scientists.

Audience: You mentioned a city on Mars named Tanio; is this a personal experience that you had, that is to say, that you saw with your own eyes?

Samael Aun Weor: Certainly, Tanio is a city that exists, and when I mention Tanio, it relates to a direct report received by a man who belongs to a group of people who travelled to Mars.

It is good for you to know that at this moment the disciples of Marconi continue with their scientific research within the jungles of the Amazon. They were performing their experiments when a group of planetary ships came from Mars and visited their installations in the Amazon. Ever since, they have been in contact with the inhabitants of Mars. They do not have contact with modern civilization anymore. Indubitably, they learned from the Martians how to utilize solar energy. The last exploration that they did was to the city of Tanio.

Since I have been in contact with the scientist who went to Mars and who visited Tanio, I do not have any inconvenience to declare it here before the audience, in order for you all to know that in detail. Likewise, I know one of the scientists of the Amazon, who nowadays is on Mars and is married with a Martian woman.

The inhabitants of Mars want friendly relations with Earthlings; they want a Mars-Earth alliance, which would be beneficial. But they are waiting for the big wave of destruction of the inhabitants of the Earth to pass by, because I repeat, at these moments the Earthlings are very dangerous. Human beings from Mars know that after the catastrophe, they will be able to establish the Mars-Earth alliance with the new race.

Mars is our neighboring planet and has a great civilization that want good relations with the Earthlings. Mr. Tage, an inhabitant of Mars who always descends in the jungles of the Amazon, has become a paladin of that great alliance; I am sure that it will be realized. But first of all, it is necessary to select the inhabitants of the planet Earth.

Audience: Why are the Blue Men waiting for the catastrophe to happen? Why are they waiting, and do not begin once and for all to help people? Why are they going to do it in an absurd manner? I think that if they are going to mix themselves with the people and take them one by one, and since they have much power, why do not they do it in intelligent manner? But only helping one, two, or three people who find in their way, seeing that they can help more. What are they waiting for?

Samael Aun Weor: The question is quite interesting; obviously it is the best question that has come today from this great audience; the one who asked reflects much intelligence and great courage.

Indeed, truly, if the blue race appeared now in cities like New York, Paris, or London, I am sure that they would not be very well treated. What fate would await the extraterrestrials if they became visible and tangible in the great capitals of the world? What would become of their ships? Are you sure that the multitudes would respect their ships? Are you sure that the multitudes would respect the lives of the inhabitants of other planets? Until now, the facts have demonstrated the opposite.

When two ships flew through the skies of the United States of North America, they wanted indeed what the gentleman has mentioned: contact with humanity. But, the North American government prepared to welcome them with bullets, machine guns, and missiles. Yes, that is the type of reception that we give to the humanities of other worlds.

If Earthlings could capture the extraterrestrials alive, they would take them to the laboratories in order to study them and to know of what are they made of, if they are made of paste or starch. And if their ships could be captured, they would be loaded with atomic weapons in order to destroy defenseless cities, and to dominate other planetary humanities by force.

Indeed, we are not meek sheep. I have known people who say, "Well, why are those inhabitants of other worlds, those extraterrestrials, flying around... it would be good if we shoot them down first, and clear our doubts once and for all."

The extraterrestrials know very well the fate that awaits them, that is, the Earthlings would assassinate them. Many prophets have been sent to the Earth,

and Earthlings have assassinated them;
many Avatars were sent in the old times,
and some of them were poisoned, others
stabbed, others sent to exile, others were
hanged. Earthlings poisoned Buddha
himself. Other wise men were sent to the
Earth, and the Earthlings terminated
them also. Finally, the Rabbi, the divine
Rabbi of Galilee, was sent to the Earth,
and the Earthlings also crucified him...
Thus, this is how Earthlings are.

The Antichrist of false science performs
inconceivable efforts in order to hide anything that in one way or another tries to
eclipse him. Such is the extraordinary case
that happened on the hill of the Three
Crosses, in the town of Cruz Alta and
Villa Real, state of Puebla, Mexico—a place
that in order to reach it, one needs to
walk several hours, given the complicated
state of the land.

Due to certain mechanical faults, a
spaceship twenty-five meters in diameter
needed to descend in the mentioned geographic place. The ship was manned by
four extraterrestrials. One of them died
and one was wounded, who was properly
taken care of by medical science. The

other two, who were unharmed, were captured and taken to the United States.

It is not irrelevant to emphatically affirm that the precious ship was taken to the United States with shameful goals and intentions. For several days the Army mounted a ring of security around the ship. The two crew members who were unharmed were seen by the people; they had an average stature of 2.5 meters [8.2 feet], their suits were of an immaculate whiteness and their silence absolute, since they did not talk with any person who approached them.

We are talking about concrete, clear, and definitive facts. Obviously, the Antichristic forces fight to gag the people; they do not want people to know the truth. They mortally hate anything that in one or any other way demonstrates the falsity of materialistic science.

I declare that at these moments we do not know the fate that the two survivors had in the hands of the Earthlings. Likewise, we do not know more about the fate of the wounded one, and we are unaware if the corpse of the one who died

was buried or taken to a special laboratory.

With this unfortunate accident, it has
been demonstrated again to satiation
the true existence of the cosmic ships,
a theme that your friend, the author of
these writings, initiated in the year 1950,
and that since then was rejected by the
learned ignoramuses of all times. I have
the high honor of being the first person
to speak about the extraterrestrials and
their ships. Today, with this unusual and
tragic event, everything that I have said
from the year 1950 is properly verified.

I do not care about the retaliation that
may be unleashed against my persona,
because what is of interest to me is to
present the truth to the world no matter
what the cost might be...

Chapter 14

The Consequences of the Comet Kondoor

First of fall, it is necessary to comprehend the meaning of our studies and their fundamental objective.

Indeed, we need to leave the condition in which we find ourselves, since this terrestrial humanity is abnormal, unbalanced; we need to comprehend this.

The psyche of the intellectual animal—mistakenly called "human"—is altered.

As a consequence or corollary of all of this, we can state that the entire terrestrial humanity is mentally, psychically, and volitionally unbalanced.

I want you to understand that during the times of the Lemurian continent, when humanity still did not possess the abominable Kundabuffer organ, all people were balanced, and they lived in harmony and peace. Yet, an unexpected event occurred as a result of the erroneous calculations of certain sacred individuals: the comet Kondoor collided with the planet Earth. Then, as consequence of that collision, a frightful catastrophe was produced.

Some gigantic islands—better said, quasi-continents—that were densely populated, sank into the ocean, and millions of human beings perished. The Earth was in a chaotic state, leaving the geological layers in an unstable situation. Therefore, the layers trembled incessantly. Great earthquakes and frightful tsunamis were produced, and since the equilibrium of the geological layers was lost, there was no stability in our world and human life threatened to disappear.

During that age, some sacred individuals came to the Earth. I want to emphatically mention the Archangel Sakaki and his most high commission, consisting of a group of specialists. These individuals studied the problem and decided that in order to stabilize the geological layers of our world, it was necessary to implant the abominable Kundabuffer organ in our humanity.

It is obvious that the physical body is a machine that captures specific types of cosmic energies, that are soon transformed and retransmitted to the interior layers of the Earth. Yes, the Earth is nourished, because the Earth is a living organism that needs nourishment, nutrition.

Therefore, when that alteration to the human organisms was allowed, the cosmic forces were also altered and became lunar. It so happened then, that from the coccyx—the fundamental bone of the dorsal spinal—the sacred fire was precipitated downward, and in its turn such a precipitated fire developed the inferior part of the dorsal spine, thus causing to grow from the coccyx (the base of the spinal column) an appendage similar to the tail that we see in the simians. Tailed organisms are fundamentally lunar. So, the

The coccyx

lunar forces stabilized the geological layers of our planet Earth.

Nevertheless, a certain miscalculation happened in the procedures of the Archangel Sakaki and his most high commission, when they allowed the Kundabuffer organ to be kept in the human being beyond the normal, objective time calculations. Thus, time passed, and during later periods when the Chief-Common-Universal-Arch-Physicist Seraphim-Angel Looisos came to the Earth, he became aware of those miscalculations.

Since these sacred individuals know how to maneuver the cosmic energies and have power over life and death, by maneuvering the forces of the cosmos, they in fact managed to begin the process of terminating the Kundabuffer organ, thus managing to destroy that organ totally.

Even still, until this day, a "little bone" exists at the base or inferior part of the dorsal spine that is even known in medicine as "the tail," which as a diminutive they call it "little tail." Thus, indeed, truly, a residue of the abominable Kundabuffer organ still remains in the anatomy of the human body.

Once the Kundabuffer organ was eliminated, it could be clearly verified by close observation that the abominable consequences of that organ remained within the five cylinders [or centers] of the organic machine. These five cylinders are:

Intellectual (situated in the brain)

Emotional (situate in the heart)

Motor (superior part of the dorsal spine)

Instinctual (inferior part of the dorsal spine)

Sexual (genitalia)

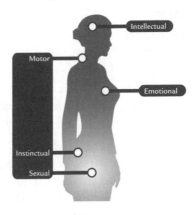

The evil consequences of the abominable Kundabuffer organ remained within these five aforementioned centers.

There are two more centers in the human being; they are of a superior type: the superior emotional and superior mental. These were not affected.

Nonetheless, the harmful psychic elements remained deposited within the machine. These elements are the psychic aggregates; in other words we would say: the "I"s. In Egypt, they were called "the red demons of Seth."

Anyhow, such aggregates, even when they are invisible to the physical eyes, are visible to the sense of psychological self-observation.

Indubitably, the Essence—that is, the consciousness of the Being—was bottled up or encapsulated within such aggregates, living personification of our psychological types of defects, namely: anger, greed, lust, envy, pride, laziness, gluttony, etc.

Regrettably, over time, these aggregates became stronger and stronger. Thus, indeed, nowadays, the Essence of the human being, which is the enlivening, the most decent element that we have within our interior, is conditioned by the psychic aggregates, since it is bottled up.

There is something more that is bottled up: the mental or "manasic Essence," and our will, etc.

Thus, in these conditions, each one of us is a multiplicity, not an individuality. This is why in the Christic gospel of Mark, Jesus of Nazareth asked the man with an unclean spirit:

> *What is thy name? And he answered, saying, My name is Legion: for we are many.* - Mark 5:9

Thus, each one of us is Legion.

It is also stated in the gospel of Luke:

> *And it came to pass afterward, that he (the great Kabir Yeshua ben Pander) went throughout every city and village, preaching and shewing the glad tidings of the kingdom of God: and the twelve were with him, and certain women, which had been healed of evil spirits and infirmities, Mary called Magdalene, out of whom went seven devils.* - Luke 8:1, 2

These seven devils inevitably personify the seven capital sins. There is no doubt that these seven multiply into many evil spirits, yet in the biblical gospel only the main seven are cited.

Nevertheless, we must also remember Virgil, the poet from Mantua, when he said:

> *No, not if I had a hundred mouths, a hundred tongues, and throats of brass, inspired with iron lungs, I could not half those my horrid crimes describe, nor half the punishments those crimes have met.*
>
> - Virgil, *The Aeneid*, book six

Those were the words of the Master of the Florentine Dante; they invite us to reflection.

Therefore, having the consciousness, the will, and even the mental essence bottled up within the aggregates, inevitably we are conditioned in the consciousness, in the will, and even in the mental psychic field. For such cause I emphatically say that this terrestrial humanity is imbalanced.

We know well that the red demons of Seth that personify our many selves quarrel amongst themselves; they do not have concordance or harmony whatsoever. When one of them emerges, he imposes himself to achieve control of the organic machine; the rest of them fight for supremacy. Finally, the one who for some moments dominates the brain, soon falls, leaving the place to another.

When looking at these things from all of these angles, we can evidence that we do not have a single mind, but thousands of minds that contradict, that discuss amongst themselves. Thus, we do not have a defined will. A permanent ego does not exist within us, but thousands of wills.

We can evidence that our Essence is turned asunder into pieces, shaken by the storms of inferior emotions; in a word, we are abnormal creatures.

I invite you to think a little. Let us observe a jealous man: how can we call him sane when his jealousy transforms a flea into a horse? Yet if the woman is the jealous one, she cannot see with tranquility her man next to another woman, or him conversing very quietly with her; and vice versa, since the same happens with the males. Behold the attitudes that the jealous ones take, what quarrels! Those who are very jealous suffer because of a simple glance; they are completely demented.

Let us now observe a person filled with hatred, which is monstrous and abominable. He hates the entire world, he does not love anyone, he abhor everybody, and makes everybody abhor him. He is completely demented, and all of his actions are madness; there is no harmony within him.

Let us observe an angry one, thundering with thunderbolts, striking others with his hands and feet, with his eyes out their sockets; he is completely demented.

Let us examine the lustful ones, how abominable they turn out to be: what perverted attitudes they take, what filthy glances; they are abnormal.

Let us observe the avaricious ones: their disarranged physiognomy, the eyes of the avaricious ones are unmistakable, their actions, their proceeding manners, namely, hiding the money and even suffering from hunger since they do not want to spend; to that end, they are crazy, demented ones.

Therefore, truly, I tell you that the people of the planet Earth are demented, mentally imbalanced. What is worse is that they are not aware that they are mentally imbalanced. They believe they are using their mental faculties perfectly. They can only become aware of their mental imbalance the day they are no longer mentally imbalanced.

Nobody can become perfectly mentally equilibrated so long as he continues with the ego, the "I," alive.

Therefore, equilibrium is achieved by eliminating all of those psychic aggregates that in their conjunction constitutes the ego, the "I", the myself, the self-willed.

There is no doubt that the people of the Earth would not even remotely accept that they are mentally imbalanced; nevertheless, they are.

Was the First World War a matter for sane people?

Was the Second World War a matter for sane people?

Was the launching of the atomic bomb over Hiroshima and Nagasaki a matter for sane people? Listen, only a mad person could do that. Only demented people would dare to do that. Yes: mentally imbalanced people.

Thus, indeed, truly, we propose to eliminate those psychic aggregates that we carry within our interior, and which make us abnormal people.

How can we achieve this; through what methods?

It is obvious that psychological self-observation is necessary in order to achieve it. Thus, when one admits that one possesses their particular, individual psychology, in fact, one begins to self-observe; and when one self-observes, one is discovering oneself, and self–revelation exists in every self-discovery.

Nevertheless, we need to perform psychological self-observation in a continuous manner, from instant to instant, from moment to moment.

The defects that we carry within flourish while in relation with our acquaintances, in the factory, in the country, in our house, and if we self-observe, we see them.

A discovered defect must be submitted to the inner self-reflection of the Being.

It is urgent to comprehend them in all of the levels of the mind.

Thus, when a defect has been properly comprehended, we need to eliminate it.

We must apprehend the complete differentiation between comprehension and elimination. To comprehend is not enough; we need to eliminate. I.e. Anyone can comprehend that he has the defect of lust, yet this does not signify that he has eliminated it. Anyone can comprehend that he has the defect of revenge, yet this does not signify that he has eliminated it.

The mind by itself cannot alter any defect. A power superior to the mind is necessary; fortunately, such a power exists in a latent manner within our interior. I want to emphatically refer to Devi Kundalini Shakti, the divine Cosmic Mother; she is a variant of our Being. She is our Being, but a derivative, a derivation; I learned this precisely from Her, Herself.

On a certain occasion in Tibet, I interrogated my Divine Mother Kundalini as follows, "You and I look like two different Beings; nonetheless, we are the same Being."

Answer, "Yes, my son, we are the same Being, but derivatives."

Thus, this must be understood. She is Marah, Isis, Adonia, Insobertha, Rhea, Cybeles, Astarte, Diana, Tonantzin, etc., She is the igneous serpent of our magical powers. Thus, when the devotee invokes Her, She then assists him. Obviously, She has the power to eliminate any given psychological defect.

Undoubtedly, those who work in the flaming forge of Vulcan can invoke her precisely in those moments in which the lingam-yoni of the Greek mysteries are found properly connected. She, reinforced by the transcendental sexual electricity, rapidly disintegrates any given aggregate that has been previously comprehended.

Those who do not yet work in the forge of the Cyclops can invoke Her during their meditation, and She will assist them by disintegrating their psychological aggregates.

I clarify: one can eliminate 100% of one's psychic aggregates in the work that one performs in the forge of the Cyclops; yet, when one is not working in the flaming forge of Vulcan, one can eliminate 25 or 30% of the aggregates.

So, by all means, without this annular serpentine power that develops in the body of the Gnostic ascetic, the absolute pulverization of the undesirable psychic elements that we carry within would not be possible.

What I am talking about is transcendental; we need to work on ourselves if we truly want to change and become normal individuals.

It is necessary to understand that the disintegration of the undesirable elements is always very difficult. Regardless, those sacred individuals who made such a mistake because of their erroneous calculations, indeed inflicted a great harm upon us, thus they have to pay their debts in future Mahamanvantaras in accordance with the law of Nemesis.

It is necessary to understand that when the psychic aggregates appeared within our organic nature as an outcome of the abominable Kundabuffer organ, they were processed within our psyche in accordance with the Law of Seven. Due to this, the total disintegration of all the undesirable psychic elements that we carry within our interior is frightfully difficult.

Let us take into account, in order to be more clear, that these elements process themselves in seven levels of the Being. Some saints achieved the elimination of aggregates in two or three or four levels, or even five, yet those who have achieved the elimination of those aggregates in all seven levels of the Being are very rare.

Therefore, we face a very difficult work. If you believe that it is possible to achieve final liberation by means of another way, listen: you are absolutely wrong.

I am widely illustrating this theme for you so that you can understand that these studies are the reason for us to be gathered here. Obviously, we need to know our purpose in gathering ourselves here, in these studies, and for what. If curiosity is the simple motive that moves you, listen: there are many things to be curious about, i.e. in city entertainment centers, in cinemas, the bulls in the arena, etc. Yet, to enter into these studies is something very serious, because to reduce the psychic aggregates to cosmic dust within the seven levels of the Being is not a very simple task. Indeed, to emancipate the Essence, to disembottle the mind and will, is not an easy task.

The mind in itself is a substance, it is the manas, yet it is bottled up within the aggregates, and therefore, it has become not one mind but many minds, thousands of minds.

Each psychic aggregate has its own mind, and since we have thousands of aggregates, then thousands are the minds that we have. Truly, we have 10,000 psychic aggregates that we need to turn into dust and which process themselves in seven levels.

In esotericism, the psychic aggregates are denominated "whales."

In the Old Testament in the Bible, it is stated,

> *Saul hath slain his thousands, and David his ten thousands.* - 2 Samuel 18:7

It is necessary to know how to understand this.

A short time ago I visited a Mayan palace in Cancun that has 10,000 crosses of Saint Andrew. So, let it be understood that to pulverize 10,000 aggregates is not an easy task. Thus, if you are willing to perform this task, I congratulate you. Yet, if you are not willing, listen: the submerged devolution within the bowels of the Earth is waiting for you, because if you cannot do it, the Earth can. Yes, if you are not capable, the laws of nature are capable. If you do not do the work, the laws of nature will do it for you down there, within the infernal worlds.

I do not advise you to devolve within the bowels of the submerged mineral kingdom, since the laws are frightfully multiplied there, thus the sufferings are terrible, until the Second Death. Thereafter, at the last instant the Essence—having suffered a lot—becomes free, and it emerges again to the surface of the Earth in order to initiate new evolving processes.

It is preferable that you perform the work here and now, because now you are before a dilemma: whether you do it or not. If you do it, marvelous, you will then be free; yet if you do not do it, nature will take care of such a work within the bowels of the world. Such is the crude reality of the facts.

Nevertheless, I know the Earthlings very well. Thus, you might think that these are just conceptions of my intellect, yet truly I tell you that this is not a matter of lucubrations of the mind. I am talking about facts. Since I am an awakened individual due to the fact of having reduced my ego to cosmic dust, it is obvious that I know about the submerged devolution within the bowels of the Earth.

So, it is preferable to work upon ourselves here and now. I talk to you about facts. I talk to you about what I have seen and heard, about what is factual in a direct manner to me.

Reflect; we are gathered here in order to study. You have come here in order to listen and I have come to talk to you in a frank manner.

We need to become serious, otherwise failure awaits us. Regrettably, it is very difficult to find serious people, since the great majority are playing. Today they are in a little school and tomorrow in another, and after tomorrow in another, and likewise they live playing. They are not serious.

If you think that you can find something outside of yourselves, listen: you are mistaken. The one who does not find the truth within himself will never ever find it outside of himself.

Thus reflect, become serious, work upon yourselves, transform yourselves, because this is what is fundamental.

Inverential peace,

Samael Aun Weor

Epilogue

Open Letter

Eminencies, President of the United States of America, and Prime Minister of the Soviet Union,

Please excuse us for not mentioning your full names; we do not know in what year this letter will arrive to your hands, and as it is clearly logical, times change and we do not know who at that date will occupy the top position of your respective countries.

The intention of this open letter is to inform you that the conquest of space has already been reached in Latin America. It is clearly natural that you will smile with skepticism before such impudent information, or possibly you will consider it as "insolent." We just want to fulfill our duty, thus we advise you not to spend more money on space rockets; such money must be utilized better. Space rockets are good for nothing, since they are a true failure.

At this moment, in a secret place of South America within the deep heart of the jungle, there is a scientific society of ninety-eight eminent scientists from different European countries. This society follows the footsteps of the great wise scientist Guillermo Marconi, who with great mastery learned how to use the powerful solar energy.

At this moment, under the guidance of Martian wise men, this society is building wonderful interplanetary ships with which they have not only thoroughly studied all of your territories, but moreover they have managed to travel to the Moon and Mars.

It is not irrelevant to clarify unto you that said scientific society has enough money in order to continue its works, thanks to the economic support of the Martians.

I clarify: it would not be strange to us if you indignantly discard this letter, since your pride and skepticism are very well known on the planet Earth. Nevertheless, within a few years you will have concrete proof of our affirmations.

Listen, the liquid fuel that you use for your space rockets is ineffective for interplanetary navigation. The spaceships designed by the Martians and built by the wise people of the mentioned scientific society—under the direction of Martian wise men—are propelled by solar energy.

The scientists of this society are eminently religious people; there is even a priest among them (it does not matter which religion). We are absolutely convinced that the conquest of outer space is absolutely impossible if we exclude religiousness.

All of the inhabitants of the cosmos are profoundly religious. All of them know very well that the divine is latently and immanently found in every atom of the infinite.

The mentioned scientific society has constructed a great underground laboratory in the heart of the jungle. This laboratory has all that is necessary for the investigation.

The contact with Martians was achieved December 16 of the year 1955 at five o'clock in the afternoon. At that hour, five Martian space machines flew over the jungle, and one landed. Four Martian people descended; the Martian head of the expedition among them. Since then, contact has been established, and the Martian spaceships continue to land in that region.

The ninety-eight scientists who reside within that deep jungle of South America coexist with the Martians and are learning the science of interplanetary navigation from them.

The mentioned scientists have received from the hands of the Martian captain— His Eminence Mr. Tage—a gold sheet with the following inscription:

> *Loga (Mars), universal brother of the immense outer space, pays homage and friendship to Dogue (the Earth), with a strong longing to unite all beings, who live in one single Spirit.*
>
> *In the infinite Spirit,*
> *For glory and peace eternal.*

We congratulate the Martian Captain Mr. Tage for his four word speech. These four words are "Sundi, Dogue, Loga, Eba," which means, "God, Earth, Mars, Alliance." With this speech and the gold sheet, the alliance between Martians and Earthlings has been sealed.

We also send our congratulations to Mr. Martinelli for the beautiful and significant ring given to Mr. Tage on October 12 of the year of 1956.

The most important cosmic event of all the centuries upon the face of the Earth—after the coming of our Lord the Christ—was performed at twelve noon. One of the most illustrious members of the mentioned scientific society, the very illustrious Mr. Narciso Genovese, stated that at that precise hour the "Columbus Expedition" left for the planet Mars.

We owe very much to Mr. Narciso Genovese regarding this subject, that is, the scientific expedition to the planet Mars. Thus, if this letter arrives to him, let him receive our congratulations.

As Columbus arrived at America with three ships, likewise three cosmic ships were constructed by the terrestrial scientists under the direction of the Martians.

The names of the three cosmic ships are "Loga," "Dogue," "Eba" (Mars, Earth, Alliance).

The interior of the ships was adorned with the image of the Christ, thus, the trip was performed with total success.

The convoy was formed by the three terrestrial cosmic ships, and six Martians who accomplished the mission of escorting the terrestrial ships. Nine people formed the crew of the terrestrial ships, three for each of the terrestrial ships.

The first stage of the cosmic flight reached the Moon, thus it was absolutely verified over and over again that the moon is a dead world. The expeditionaries rested on the Moon and soon continued their trip to Mars. Ten more ships of Martian origin joined the expedition Aries on the Moon.

All the inhabitants of the city of Tanio, capital of the planet Mars, went to the port to welcome the inhabitants of the Earth.

The expeditionaries remained five days on the planet Mars, dedicated to observation and study. They learned plenty on Mars, and after their victorious return they continued in the heart of the South American jungle with their studies and investigations.

The ninety-eight European scientists dedicated to these investigations and studies under the direction of the Martian wise men, want to share their knowledge with all the inhabitants of the Earth. They want all of humanity to participate in the interplanetary navigation. Regrettably, Russia and the United States—with their atomic experiments and nuclear blasts—are hindering, impeding the members of such an august scientific society, from making us, all of the inhabitants of the Earth, participants in these cosmic trips.

The two great World Wars that filled the world with pain, and now the Cold War with all its possibilities of becoming hot, filling the world again with blood and destruction, are the main factors that impede the cultural exchange with the Martians and cosmic trips.

Space rockets are no longer necessary; the contact with the Martians is already established. Now what is necessary in order to participate in the cosmic trips is the dissolution of the "I."

As long as the "I" continues to exist, there will be no peace; and while there is no peace, interplanetary trips are impossible. In these precise moments in which we live, we do not need space rockets, but the study of the 'I' and its total death;

thus and only thus, the trips to Mars will become possible.

It is impossible to take to Mars assassins, the greedy, drunkards, gluttons, materialists, Marxists, enemies of the Eternal One, thieves, prostitutes, etc.

On Mars, peace reigns; governments, nationalities, armies, and police are not even necessary there. There are no delinquents on Mars, and if someone were born like that, he would be considered ill and would be taken to an isolated sanatorium.

Gentlemen, think about this, on what this means. Think about a world like this, a world where the "I" no longer exists. Imagine for a moment an army of the Earth invading Mars. Comprehend what such a horror, such barbarism, signifies.

The author of this letter begs you gentlemen, in the name of the truth, to end your atomic explosions, to finish with the Cold War, and to initiate a time of universal religiousness.

In a very special manner I ask the comrades of the Soviet Union to suspend the public and private diffusion of materialist dialectics, and to intensify the propaganda in favor of religion.

You must know, sirs, that all of the inhabitants of the cosmos render cult to the divinity, and that the conquest of outer space is impossible without religiousness.

Please gentlemen, I beg you in the name of the inhabitants of the Earth, do not harm us anymore with your wars, hatred of divinity, nuclear blasts, etc.

Signed in Mexico, April 29, after
the third year of Aquarius (1965),
by the president and founder
of the Gnostic Movement,

Samael Aun Weor

Glossary

Absolute: Abstract space; that which is without attributes or limitations. Also known as sunyata, void, emptiness, Parabrahman, Adi-buddha, and many other names.

The Absolute has three aspects: the Ain, the Ain Soph, and the Ain Soph Aur.

"The Absolute is the Being of all Beings. The Absolute is that which Is, which always has Been, and which always will Be. The Absolute is expressed as Absolute Abstract Movement and Repose. The Absolute is the cause of Spirit and of Matter, but It is neither Spirit nor Matter. The Absolute is beyond the mind; the mind cannot understand It. Therefore, we have to intuitively understand Its nature." - Samael Aun Weor, *Tarot and Kabbalah*

"In the Absolute we go beyond karma and the gods, beyond the law. The mind and the individual consciousness are only good for mortifying our lives. In the Absolute we do not have an individual mind or individual consciousness; there, we are the unconditioned, free and absolutely happy Being. The Absolute is life free in its movement, without conditions, limitless, without the mortifying fear of the law, life beyond spirit and matter, beyond karma and suffering, beyond thought, word and action, beyond silence and sound, beyond forms." - Samael Aun Weor, *The Major Mysteries*

Archangel: (Greek) From arch, chief or primordial, and angel, messenger.

Aryan Race: Quoted from *Webster's Revised Unabridged Dictionary:* "From Sanskrit *arya,* for excellent, honorable; akin to the name of the country Iran, and perhaps to Erin, Ireland, and the early name of this people, at least in Asia. One of a primitive people supposed to have lived in prehistoric times, in Central Asia, east of the Caspian Sea, and north of the Hindoo Koosh and Paropamisan Mountains, and to have been the stock from which sprang the Hindoo, Persian, Greek, Latin, Celtic, Teutonic, Slavonic, and other races; one of that ethnological division of mankind called also Indo-European or Indo-Germanic."

In Universal Gnosticism, the Aryan root race refers to the vast majority of the popluation of this planet, and is noted for its close relationship with Ares or Mars, the God of War. Compare with AquAryan and barbAryan. The Aryan race, the fifth great race to exist on this planet, is under the guidance of Ares, Mars, the Fifth of the Seven Spirits before the Throne of God.

"Every Root Race has seven Subraces. The seed of our Aryan Root Race is Nordic, but when the Nordics mixed themselves with the Atlantean survivors, they gave origin onto the Subraces of the Aryan trunk.

"First Subrace: It flourished in central Asia, in those now vanished kingdoms of central Asia, and whose ruins still exist in the Himalayas around the country of Tibet. Powerful spiritual civilizations of the first Aryan Subrace existed in those regions.

"Second Subrace: It flourished in India and the entire south of Asia. In Pearland, the sacred land of the Vedas, in the ancient Hindustan, where the second Aryan Subrace developed, formidable esoteric cultures and tremendous civilizations existed.

"Third Subrace: It created powerful civilizations. Babylon, Chaldea, Egypt, etc., etc. were the scenario of very rich and powerful civilizations created by the third Aryan Subrace.

"Fourth Subrace: It developed in Rome, Greece, Italy, and Athens, the great city founded by the Goddess Athena. Before their degeneration and destruction, Greece and Italy were marvelous scenarios where the powerful civilizations of the fourth Aryan Subrace developed.

"Fifth Subrace: Are the Anglo-Saxon and Teutonic. The First and Second World Wars, with all of their barbarism and moral corruption, point with their accusatory fingers to the men and women of the fifth Aryan Subrace.

"Sixth Subrace: The mixture of the Spanish Conquistadors with the Native-American tribes. The effort to form the sixth Subrace in the redskin territory was very difficult, because the English Conquistadors destroyed them; they assassinated them, instead of mixing themselves with the natives. Only in a very insignificant and incipient way was the mixture of blood performed. This is why the Occult Fraternity saw the necessity of converting the North American territory into a melting crucible of races. So, the formation of the sixth Subrace in the United States had enormous difficulties; there, all the races of the world have mixed. The sixth Subrace in Latin America was formed very easily and this is something that must not be ignored by the treatisers of anthropogenesis and occultism.

"Seventh Subrace: The survivors of the new great cataclysm that soon will destroy this Aryan Root Race will be formed by the survivors of the Seventh Subrace; they still do not exist, but they will.

"So, this Aryan Root Race, instead of evolving, has devolved, and its corruption is now worse than that of the Atlanteans in their epoch. Its wickedness is so great that it has reached unto heaven." - Samael Aun Weor, *The Kabbalah of the Mayan Mysteries*

Astral Body: The body utilized by the consciousness in the fifth dimension or world of dreams. What is commonly called the Astral Body is not the true Astral Body, it is rather the Lunar Protoplasmatic Body, also known as the Kama Rupa (Sanskrit, "body of desires") or "dream body" (Tibetan rmi-lam-gyi lus). The true Astral Body is Solar (being superior to Lunar Nature) and must be created, as the Master Jesus indicated in the Gospel of John 3:5-6, "Except a man be born of water and of the Spirit, he cannot enter into the kingdom of God. That which is born of the flesh is flesh; and that which is born of the Spirit is spirit." The Solar Astral Body is created as a result of the Third Initiation of Major Mysteries (Serpents of Fire), and is perfected in the Third Serpent of Light. In Tibetan Buddhism, the Solar Astral Body is known as the illusory body (sgyu-lus). This body is related to the emotional center and to the sephirah Hod.

"Really, only those who have worked with the Maithuna (White Tantra) for many years can possess the Astral Body." - Samael Aun Weor, *The Elimination of Satan's Tail*

Being: Our inner, divine Source, also called the Innermost or Monad, which is not easily definable in conceptual terms. The use of the term "Being" is important though, in relation to its roots, as shown in the Online Etymology Dictionary: "O.E. beon, beom, bion "be, exist, come to be, become," from P.Gmc. *beo-, *beu-. This "b-root" is from PIE base *bheu-, *bhu- "grow, come into being, become," and in addition to the words in English it yielded German present first and second person sing. (bin, bist, from O.H.G. bim "I am," bist "thou art"), L. perf. tenses of esse (fui "I was," etc.), O.C.S. byti "be," Gk. phu- "become," O.Ir. bi'u "I am," Lith. bu'ti "to be," Rus. byt' "to be," etc. It also is behind Skt. bhavah "becoming," bhavati "becomes, happens," bhumih "earth, world."

Centers, Seven: The human being has seven centers of psychological activity. The first five are the Intellectual, Emotional, Motor, Instinctive, and Sexual Centers. However, through inner development one learns how to utilize the Superior Emotional and Superior Intellectual Centers. Most people do not use these two at all.

Christ: Derived from the Greek Christos, "the Anointed One," and Krestos, whose esoteric meaning is "fire." The word Christ is a title, not a personal name. Christ is a universal force, known by many names, such as Quetzalcoatl, Avalokitesvara, Krishna, Jupiter, Osiris, and many more.

Consciousness: "Wherever there is life, there is consciousness. Consciousness is inherent to life as humidity is inherent to water." - Samael Aun Weor, *Fundamental Notions of Endocrinology and Criminology*

From various dictionaries: 1. The state of being conscious; knowledge of one's own existence, condition, sensations, mental operations, acts, etc. 2. Immediate knowledge or perception of the presence of any object, state, or sensation. 3. An alert cognitive state in which you are aware of yourself and your situation. In universal Gnosticism, the range of potential consciousness is allegorized in the Ladder of Jacob, upon which the angels ascend and descend. Thus there are higher and lower levels of consciousness, from the level of demons at the bottom, to highly realized angels in the heights.

"It is vital to understand and develop the conviction that consciousness has the potential to increase to an infinite degree." - The 14th Dalai Lama

"Light and consciousness are two phenomena of the same thing; to a lesser degree of consciousness, corresponds a lesser degree of light; to a greater degree of consciousness, a greater degree of light." - Samael Aun Weor, *The Esoteric Treatise of Hermetic Astrology*

Demiurge: (Greek, for "worker" or "craftsman") The Demiurgos or Artificer; the supernal power that built the universe. Freemasons derive from this word their phrase "Supreme Architect." Also the name given by Plato in a passage in *Timaeus* to the creator God.

"Esotericism admits the existence of a Logos, or a collective Creator of the universe, a Demiurge architect. It is unquestionable that such a Demiurge is not a personal deity as many mistakenly suppose, but rather a host of Dhyan Chohans, Angels, Archangels, and other forces." - Samael Aun Weor, *The Three Mountains*

"It is impossible to symbolize or allegorize the Unknowable One. Nevertheless, the Manifested One, the Knowable Elohim, can be allegorized or symbolized. The Manifested Elohim is constituted by the Demiurge Creator of the Universe. [...] The great invisible Forefather is Aelohim, the Unknowable Divinity. The great Triple-Powered God is the Demiurge Creator of the Universe: Multiple Perfect Unity. The Creator Logos is the Holy Triamatzikamno. The Verb, the Great Word. The three spaces of the First Mystery are the regions of the Demiurge Creator." - Samael Aun Weor, *The Pistis Sophia Unveiled*

"The Demiurge Architect of the Universe is not a human or divine individual; rather, it is Multiple Perfect Unity, the Platonic Logos." - Samael Aun Weor, *Gnostic Anthropology*

Devolution: (Latin) From devolvere: backwards evolution, degeneration. The natural mechanical inclination for all matter and energy in nature to return towards their state of inert uniformity. Related to the Arcanum Ten: Retribution, the Wheel of Samsara. Devolution is the inverse process of evolution. As evolution is the complication of matter or energy, devolution is the slow process of nature to simplify matter or energy by applying forces to it.

Diogenes: A controversial Greek philosopher who extolled poverty as virtuous, begged for a living, and slept in a tub in the marketplace. He was notorious for his provocative and confrontational behavior, and was one of the only people to survive publicly insulting Alexander, who deeply respected Diogenes. One story relates that Diogenes walked the city carrying a lamp in the daytime, looking for an true man but not finding any.

Ego: The multiplicity of contradictory psychological elements that we have inside are in their sum the "ego." Each one is also called "an ego" or an "I." Every ego is a psychological defect which produces suffering. The ego is three (related to our Three Brains or three centers of psychological processing), seven (capital sins), and legion (in their infinite variations).

"The ego is the root of ignorance and pain." - Samael Aun Weor, *The Esoteric Treatise of Hermetic Astrology*

"The Being and the ego are incompatible. The Being and the ego are like water and oil. They can never be mixed... The annihilation of the psychic aggregates (egos) can be made possible only by radically comprehending our errors through meditation and by the evident Self-reflection of the Being." - Samael Aun Weor, *The Pistis Sophia Unveiled*

Essence: See: Consciousness.

"Without question the Essence, or Consciousness, which is the same thing, sleeps deeply... The Essence in itself is very beautiful. It came from above, from the stars. Lamentably, it is smothered deep within all these "I's" we carry inside. By contrast, the Essence can retrace its steps, return to the point of origin, go back to the stars, but first it must liberate itself from its evil companions, who have trapped it within the slums of perdition. Human beings have three percent free Essence, and the other ninety-seven percent is imprisoned within the "I's"." - Samael Aun Weor, *The Great Rebellion*

"A percentage of psychic Essence is liberated when a defect is disintegrated. Thus, the psychic Essence which is bottled up within our defects will be completely liberated when we disintegrate each and every one of our false values, in other words, our defects. Thus, the radical transformation of ourselves will occur when the totality of our Essence is liberated. Then, in that precise moment, the eternal values of the Being will express themselves through us. Unquestionably, this would be marvelous not only for us, but also for all of humanity." - Samael Aun Weor, *The Revolution of the Dialectic*

Flaming Forge of Vulcan: A reference to the ancient secret knowledge of Tantra or Alchemy related to sexual transmutation.

Gnosis: (Greek) Knowledge.

1. The word Gnosis refers to the knowledge we acquire through our own experience, as opposed to knowledge that we are told or believe in. Gnosis - by whatever name in history or culture - is conscious, experiential knowledge, not merely intellectual or conceptual knowledge, belief, or theory. This term is synonymous with the Hebrew "daath" and the Sanskrit "jna."

2. The tradition that embodies the core wisdom or knowledge of humanity.

"Gnosis is the flame from which all religions sprouted, because in its depth Gnosis is religion. The word "religion" comes from the Latin word "religare," which implies "to link the Soul to God"; so Gnosis is the very pure flame from where all religions sprout, because Gnosis is Knowledge, Gnosis is Wisdom." - Samael Aun Weor, *The Esoteric Path*

"The secret science of the Sufis and of the Whirling Dervishes is within Gnosis. The secret doctrine of Buddhism and of Taoism is within Gnosis. The sacred magic of the Nordics is within Gnosis. The wisdom of Hermes, Buddha, Confucius, Mohammed and Quetzalcoatl, etc., etc., is within Gnosis. Gnosis is the Doctrine of Christ." - Samael Aun Weor, *The Revolution of Beelzebub*

God: "The love that all mystic institutions of the world feel for the divine is noticeable: for Allah, Brahma, Tao, Zen, I.A.O., INRI, God, etc. Religious esotericism does not teach atheism of any kind, except in the sense that the Sanskrit word nastika encloses: no admission of idols, including that anthropomorphic God of the ignorant populace. It would be an absurdity to believe in a celestial dictator who is seated upon a throne of tyranny and throws lightning and thunderbolts against this sad human ant hill. Esotericism admits the existence of a Logos, or a collective Creator of the universe, a Demiurge architect. It is unquestionable that such a Demiurge is not a personal deity as many mistakenly suppose, but rather a host of Dhyan Chohans, Angels, Archangels, and other forces. God is Gods. It is written with characters of fire in the resplendent book of life that God is the Army of the Voice, the great Word, the Verb. "In the beginning was the Word, and the Word was with God, and the Word was God. All things were made by him, and without him was not any thing made that was made." (John 1:1-3) For this reason, it is evident that any authentic human being who really achieves perfection enters into the Current of Sound, into the Celestial Army constituted by the Buddhas of Compassion, Angels, Planetary Spirits, Elohim, Rishi- Prajapatis, etc. It has been said to us that the Logos sounds and this is obvious. The Demiurge, the Verb, is the multiple, perfect unity. Whosoever adores the Gods, whosoever surrenders worship unto them, is more capable of capturing the deep significance of the diverse divine facets of the Demiurge architect. When humanity began to mock the Holy Gods, then it fell mortally wounded into the gross materialism of this Iron Age." - Samael Aun Weor, *The Three Mountains*

Human Being: In general, there are three types of human beings:

1. The ordinary person (called human being out of respect): the intellectual animal

2. The true Human Being or Man (from manas, mind): someone who has created the Soul (the Solar Bodies, symbolized as the chariot of Ezekiel or Krishna, the Wedding Garment of Jesus, the sacred weapons of the heroes of mythology, etc. Such persons are Saints, Masters, or Buddhas of various levels.

3. The Superman: a true Human Being who has also incarnated the Cosmic Christ, thus going beyond mere Sainthood or Buddhahood, and into the highest reaches of liberation. These are the founders of religions, the destroyers of dogmas and traditions, the great rebels of spiritual light.

According to Gnostic anthropology, a true human being is an individual who has conquered the animal nature within and has created the Soul (the Mercabah of the Kabbalists, the Sahu of the Egyptians, the To Soma Heliakon of the Greeks): this is "the Body of Gold of the Solar Man." A true Human Being is one with the Monad, the Inner Spirit. It can be said that the true Human Being or Man is the inner Spirit (in Kabbalah, Chesed. In Hinduism, Atman).

"Every spirit is called man, which means that only the aspect of the light of the spirit that is enclothed within the body is called man. So the body of the spirit of the holy side is only a covering; in other words, the spirit is the actual essence of man and the body is only its covering. But on the other side, the opposite applies. This is why it is written: 'you have clothed me with skin and flesh...' (Iyov 10:11). The flesh of man is only a garment covering the essence of man, which is the spirit. Everywhere it is written the flesh of man, it hints that the essence of man is inside. The flesh is only a vestment for man, a body for him, but the essence of man is the aspect of his spirit." - Zohar 1A:10:120

A true Human Being has reconquered the innocence and perfection of Eden, and has become what Adam was intended to be: a King of Nature, having power over Nature. The Intellectual Animal, however, is controlled by nature, and thus is not a true Human Being. Examples of true Human Beings are all those great saints of all ages and cultures: Jesus, Moses, Mohammed, Krishna, and many others whose names were never known by the public.

Intellectual Animal: A term describing the current state of humanity: animals with intellect.

When the Intelligent Principle, the Monad, sends its spark of consciousness into Nature, that spark, the anima, enters into manifestation as a simple mineral. Gradually, over millions of years, the anima gathers experience and evolves up the chain of life until it perfects itself in the level of the mineral kingdom. It then graduates into the plant kingdom, and subsequently into the animal kingdom. With each ascension the spark receives new capacities and higher grades of complexity. In the animal kingdom it learns procreation by ejaculation. When that animal intelligence enters into the human kingdom, it receives a new capacity: reasoning, the intellect; it is now an anima with intellect: an Intellectual Animal. That spark must then perfect itself in the human kingdom in order to become a complete and perfect Human Being, an entity that has conquered and transcended everything that belongs to the lower kingdoms. Unfortunately, very few Intellectual Animals perfect themselves; most remain enslaved by their animal nature, and thus are reabsorbed by Nature, a process belonging to the devolving side of life and called by all the great religions "Hell" or the Second Death.

"The present manlike being is not yet human;
he is merely an intellectual animal. It is a very
grave error to call the legion of the 'I' the 'soul.'
In fact, what the manlike being has is the psychic
material, the material for the soul within his
Essence, but indeed, he does not have a Soul yet."
- Samael Aun Weor, *The Revolution of the Dialectic*

"I died as a mineral and became a plant,
I died as plant and rose to animal,
I died as animal and I was Man.
Why should I fear? When was I less by dying?
Yet once more I shall die as Man, to soar
With angels blest; but even from angelhood
I must pass on: all except God doth perish.
When I have sacrificed my angel-soul,
I shall become what no mind e'er conceived.
Oh, let me not exist! for Non-existence
Proclaims in organ tones, To Him we shall
return." - Jalal al-Din Muhammad Rumi (1207 –
1273) founder of the Mevlevi order of Sufism

Kabir: The word "Kabir" is equally well-known in
India and in the Middle East, although in each
case the word means different things (in many
areas it is used as a proper name). From Arabic it
means "great" or "the Most High." It is usually
attached to the name of a great teacher, as "Sri"
is in India. Esoterically, it refers to the Divine
Calf, the child of the Divine Mother (who is symbolized by a cow in India, Greece, etc.) The word
is related to the ancient and mysterious Cabeiri
of the Greek mysteries.

Kundalini: "Kundalini, the serpent power or mystic
fire, is the primordial energy or Sakti that lies
dormant or sleeping in the Muladhara Chakra,
the centre of the body. It is called the serpen-

tine or annular power on account of serpentine form. It is an electric fiery occult power, the great pristine force which underlies all organic and inorganic matter. Kundalini is the cosmic power in individual bodies. It is not a material force like electricity, magnetism, centripetal or centrifugal force. It is a spiritual potential Sakti or cosmic power. In reality it has no form. [...] O Divine Mother Kundalini, the Divine Cosmic Energy that is hidden in men! Thou art Kali, Durga, Adisakti, Rajarajeswari, Tripurasundari, Maha-Lakshmi, Maha-Sarasvati! Thou hast put on all these names and forms. Thou hast manifested as Prana, electricity, force, magnetism, cohesion, gravitation in this universe. This whole universe rests in Thy bosom. Crores of salutations unto thee. O Mother of this world! Lead me on to open the Sushumna Nadi and take Thee along the Chakras to Sahasrara Chakra and to merge myself in Thee and Thy consort, Lord Siva. Kundalini Yoga is that Yoga which treats of Kundalini Sakti, the six centres of spiritual energy (Shat Chakras), the arousing of the sleeping Kundalini Sakti and its union with Lord Siva in Sahasrara Chakra, at the crown of the head. This is an exact science. This is also known as Laya Yoga. The six centres are pierced (Chakra Bheda) by the passing of Kundalini Sakti to the top of the head. 'Kundala' means 'coiled'. Her form is like a coiled serpent. Hence the name Kundalini." - Swami Sivananda, *Kundalini Yoga*

Law of Seven: (Also called Heptaparaparshinokh) The fundamental principle of the organization of everything that is created. Creation occurs through the Law of Three. What is created is organized by the Law of Seven. Thus we have seven primary colors, seven primary chakras, seven primary notes, seven primary planets, etc.

Law of Three: (Also called Triamazikamno) The law of nature through which action is accomplished, by means of three forces. The active principle is the positive pole, the passive principle is the negative pole, and the principle that conciliates both is the third force, the neutral force. The Law of Three is formed then by the three principles: Holy Affirmation, Holy Negation, and Holy Conciliation. The third force conciliates the affirming force with the negating force. This profound law is represented in all religions by a trinity of powers.

Lemurians: The third Root Race of this terrestrial round. The Lemurians existed before the Atlanteans, but have been confused with them by some groups. About this, H.P. Blavatsky said, "In our own day we witness the stupendous fact that such comparatively recent personages as Shakespeare and William Tell are all but denied, an attempt being made to show one to be a nom de plume, and the other a person who never existed. What wonder then, that the two powerful races -- the Lemurians and the Atlanteans -- have been merged into and identified, in time, with a few half mythical peoples, who all bore the same patronymic?" (*The Secret Doctrine,* 1888)

"It is clear that the Miocene Epoch had its proper scenario on the ancient Lemurian land, the continent that was formerly located in the Pacific Ocean. Remnants of Lemuria are still located in Oceania, in the great Australia, and on Easter Island (where some carved monoliths were found), etc." - Samael Aun Weor, *Gnostic Anthropology*

"The third Root Race was the Lemurian race, which inhabited Mu, which today is the Pacific Ocean. They perished by fire raining from the sun (volcanoes and earthquakes). This Root Race was governed by the Aztec God Tlaloc. Their reproduction was by means of gemmation. Lemuria was a very extensive continent. The Lemurians who degenerated had, afterwards, faces similar to birds; this is why some savages, when remembering tradition, adorned their heads with feathers." - Samael Aun Weor, *The Kabbalah of the Mayan Mysteries*

Manas: Sanskrit, "mind." The root of the English term "man."

Melchisedeck: Alternatively, Malkhi-tzedek, Malki Tzedek, or Melchisedek, which means "King of Justice or Virtue) See Gen. 14:18-20, Ps. 110:4, and Epistle to the Hebrews, ch. 7.

"Malkhi-tzedek king of Salem brought forth bread and wine. He was a priest to God, the Most High. He blessed [Abram], and said, 'Blessed be Abram to God Most High, Possessor of heaven and earth. And blessed be God Most High, who delivered your enemies into your hand.' [Abram then] gave him a tenth of everything." - Genesis / Bereshit 14:18-20

"[Jesus] Who in the days of his flesh, when he had offered up prayers and supplications with strong crying and tears unto him that was able to save him from death, and was heard in that he feared; Though he were a Son, yet learned he obedience by the things which he suffered; And being made perfect, he became the author of eternal salvation unto all them that obey him; Called of God an high priest after the order of Melchisedec. Of whom we have many things to say, and hard to be uttered, seeing ye are dull of hearing." - Hebrews 5

"Melchisedec is the planetary Genie of the Earth, of whom Jesus, the great Kabir, gave testimony. Melchisedec is the Great Receiver of the Cosmic Light. Melchisedec has a physical body. He is a Man, or better if we say, he is a Super-Man. The Kingdom of Agharti is found in the subterranean caverns of the Earth. The Earth is hollow and the network of caverns constitute Agharti. The Genie of the Earth lives in Agharti with a group of survivors from Lemuria and Atlantis. The Goros, powerful Lords of life and death, work with Melchisedec. The whole ancient wisdom of the centuries has been recorded on Stone, within the Kingdom of Agharti." - Samael Aun Weor, *The Pistis Sophia Unveiled*

Meditation: "When the esotericist submerges himself into meditation, what he seeks is information." - Samael Aun Weor

"It is urgent to know how to meditate in order to comprehend any psychic aggregate, or in other words, any psychological defect. It is indispensable to know how to work with all our heart and with all our soul, if we want the elimination to occur." - Samael Aun Weor, *The Pistis Sophia Unveiled*

"1. The Gnostic must first attain the ability to stop the course of his thoughts, the capacity to not think. Indeed, only the one who achieves that capacity will hear the Voice of the Silence.

"2. When the Gnostic disciple attains the capacity to not think, then he must learn to concentrate his thoughts on only one thing.

"3. The third step is correct meditation. This brings the first flashes of the new consciousness into the mind.

"4. The fourth step is contemplation, ecstasy or Samadhi. This is the state of Turiya (perfect clairvoyance). - Samael Aun Weor, *The Perfect Matrimony*

Monad: (Latin) From monas, "unity; a unit, monad." The Monad is the Being, the Innermost, our own inner Spirit.

"We must distinguish between Monads and Souls. A Monad, in other words, a Spirit, is; a Soul is acquired. Distinguish between the Monad of a world and the Soul of a world; between the Monad of a human and the Soul of a human; between the Monad of an ant and the Soul of an ant. The human organism, in final synthesis, is constituted by billions and trillions of infinitesimal Monads. There are several types and orders of primary elements of all existence, of every organism, in the manner of germs of all the phenomena of nature; we can call the latter Monads, employing the term of Leibnitz, in the absence of a more descriptive term to indicate the simplicity of the simplest existence. An atom, as a vehicle of action, corresponds to each of these genii or Monads. The Monads attract each other, combine, transform themselves, giving form to every organism, world, micro-organism, etc. Hierarchies exist among the Monads; the Inferior Monads must obey the Superior ones that is the Law. Inferior Monads belong to the Superior ones. All the trillions of Monads that animate the human organism have to obey the owner, the chief, the Principal Monad. The regulating Monad, the Primordial Monad permits the activity of all of its subordinates inside the human organism, until the time indicated by the Law of Karma." - Samael Aun Weor, *The Esoteric Treatise of Hermetic Astrology*

"(The number) one is the Monad, the Unity, Iod-Heve or Jehovah, the Father who is in secret. It is the Divine Triad that is not incarnated within a Master who has not killed the ego. He is Osiris, the same God, the Word." - Samael Aun Weor, *Tarot and Kabbalah*

"When spoken of, the Monad is referred to as Osiris. He is the one who has to Self-realize Himself... Our own particular Monad needs us and we need it. Once, while speaking with my Monad, my Monad told me, 'I am self-realizing Thee; what I am doing, I am doing for Thee.' Otherwise, why are we living? The Monad wants to Self-realize and that is why we are here. This is our objective." - Samael Aun Weor, *Tarot and Kabbalah*

"The Monads or Vital Genii are not exclusive to the physical organism; within the atoms of the Internal Bodies there are found imprisoned many orders and categories of living Monads. The existence of any physical or supersensible, Angelic or Diabolical, Solar or Lunar body, has billions and trillions of Monads as their foundation." - Samael Aun Weor, *The Esoteric Treatise of Hermetic Astrology*

Panspermia (Greek: pas/pan "all" and sperma "seed") The hypothesis that "seeds" of life exist already all over the Universe, that life on Earth may have originated through these "seeds," and that they may deliver or have delivered life to other habitable bodies.

Second Death: The complete dissolution of the ego in the infernal regions of nature, which in the end (after unimaginable quantities of suffering) purifies the Essence of all sin (karma) so that it may try again to reach complete development.

"He that overcometh (the sexual passion) shall inherit all things; and I will be his God (I will incarnate myself within him), and he shall be my son (because he is a Christified one), But the fearful (the tenebrous, cowards, unbelievers), and unbelieving, and the abominable, and murderers, and whoremongers, and sorcerers, and idolaters, and all liars, shall have their part in the lake which burneth with fire and brimstone: which is the second death. (Revelation 21) This lake which burns with fire and brimstone is the lake of carnal passion. This lake is related with the lower animal depths of the human being and its atomic region is the abyss. The tenebrous slowly disintegrate themselves within the abyss until they die. This is the second death." - Samael Aun Weor, *The Aquarian Message*

Self-observation: An exercise of attention, in which one learns to become an indifferent observer of one's own psychological process. True Self-observation is an active work of directed attention, without the interference of thought, emotion, or sensations.

"We need attention intentionally directed towards the interior of our own selves. This is not a passive attention. Indeed, dynamic attention proceeds from the side of the observer, while thoughts and emotions belong to the side which is observed." - Samael Aun Weor, *Revolutionary Psychology*

Superman: The term Superman first appeared in English in 1903 when George Bernard Shaw translated the German Übermensch, "highly evolved human being that transcends good and evil," from "Thus Spake Zarathustra" (1883-91), by Friedrich Nietzsche (1844-1900). This term refers to a type of human being far beyond the ordinary person. See: Human Being.

"The masses qualify the Superman as perverse for the very fact that he does not fit in with indisputable dogmas, neither within pious phrases, nor within the upright morality of serious people. People abhor the Superman. They crucify him amongst criminals because they do not understand him, because they prejudge him, viewing him through the psychological lenses of what is believed to be holy, even if it is evil. The Superman is like a flash of lightening which falls over the perverse, or like the brilliance of something which is not understood and which is later lost in mystery. The Superman is not a saint, nor is he perverse; he is beyond sanctity and perversity. Nevertheless, people qualify him as holy or perverse. The Superman glimmers for a moment within the dark-ness of this world and soon afterwards disappears forever. Within the Superman, the Red Christ, the revolutionary Christ, the Lord of the great rebellion radiantly shines." - Samael Aun Weor, *The Great Rebellion*

Three Brains: Gnostic psychology recognizes that humanoids actually have three centers of intelligence within: an intellectual brain, an emotional brain, and a motor/instinctive/sexual brain (they are also called "centers"). These are not physical brains; they are divisions of organized activity. Each one functions and operates independent of the others, and each one has a host of jobs and duties that only it can accomplish. Of course, in modern humanity the three brains are grossly out of balance and used incorrectly. See *Revolutionary Psychology* by Samael Aun Weor.

Index

Abduct, 11
Absolute, 36, 93, 165, 181, 197
Absolute Abstract Movement, 197
Accelerator, 82
Accident, 166
Accidents, 59, 89, 103
Achilles, 60, 145
Acrobatics, 20-21, 39
Action, 37, 136, 197, 211, 215
Actions, 36, 176
Adamski, Geroge, 54, 110
Adonia, 43, 180
Adossia, 92-93
Adulterate, 18, 27
Advanced, 13, 109, 115
Aeneid, 4, 39, 129, 174
Africa, 121
Age of Aquarius, 142
Ages, 8, 208
Aggregates, 172-174, 177-178, 180-183, 204, 214
Agriculture, 148
Ahriman, 113
Aid, 74, 94, 158
Airplane, 29, 102, 146
Airplanes, 58, 68, 84, 87
Airports, 94, 96

Albany, 57, 61
Alchemy, 205
Alcohol, 111
Alert, 42, 117, 125, 202
Alliance, 161, 190-191
Alta, 164
Altar, 120
Amazon, 121, 160-161
America, 22, 62, 136, 145, 156, 163, 187, 190-191, 199
American, 54, 101, 110, 138-139, 145, 163, 192, 199
Anahuac, 30
Analogies, 5
Analysis, 2-3, 21, 43, 110, 117
Anarchy, 59
Anatomy, 43, 149, 171
Ancestors, 30
Ancient, 77, 198, 205, 210, 212, 214
Andrew, 183
Angels, 90, 111-112, 202-203, 206, 210
Anger, 37, 118, 143, 172, 176
Anglo-Saxon, 199
Animal, 10, 63, 65, 69, 167, 207-209, 217
Animalistic, 44

Animals, 65-66, 69-70, 120, 148, 208-209

Annihilate, 43, 115-117, 204

Annular, 181, 210

Antarctic, 157

Antennas, 152

Anthropology, 155, 203, 207, 212

Anthropophagi, 86

Antichrist, 133, 164-165

Apostle Peter, 130

Apparatuses, 138, 152

Apples, 27

Aquarius, 87, 142, 195

Aramaic, 124

Archangel, 90, 92-93, 168, 170, 197, 90, 203, 206

Archangel Adossia, 92

Archangel Hariton, 92-93

Archangel Sakaki, 168, 170

Architect, 48, 202-203, 206-207

Argentina, 52, 58, 96

Aries, 192

Arizona, 139

Armaments, 35, 60

Armed, 68, 96, 129

Armies, 34, 36, 194

Arms, 35, 50, 145

Army, 134, 145, 165, 194, 206

Aroma, 13

Arrhenius, 4

Aryan, 155, 198-200

Asia, 198

Asleep, 42, 69, 147

Assassinate, 11, 66, 146, 163-164, 194, 199

Astarte, 180

Asteroids, 150, 159

Astral, 43, 85, 116, 200-201

Astral Body, 85, 200-201

Astronauts, 20, 79

Astronomers, 150

Astronomy, 124

Athena, 44, 199

Athens, 63, 199

Atlanteans, 94, 154, 198, 200, 212

Atlantic, 22

Atlantis, 77, 94, 154, 214

Atmosphere, 26, 72, 91, 104, 124-125, 127, 130-131, 148-149

Atom, 20, 25, 189, 215

Atomic, 25, 27, 30, 35-36, 73, 95, 110, 118, 124-125, 127, 129, 140-141, 150, 157, 159, 163, 177, 193-194, 217

Atoms, 32, 216

Atrocities, 73, 140

Atrophied, 41

Attention, 4, 102, 217
Attitude, 74, 175-176
Australia, 52, 212
Avatars, 151, 164
Awaken, 42, 50, 143
Awakened, 47, 49, 102, 185
Awakening, 143
Aware, 59-60, 170, 177, 202
Axes, 28, 153-154
Axis, 82, 84
Aztec, 155, 212
Babel, 79, 131
Babylon, 199
Barbarian, 74, 128, 134, 194, 199
Barranquilla, 136
Believe, 5, 7, 17-18, 24, 34, 38, 46, 48-49, 53, 70, 76, 86, 103, 133-134, 142, 151, 177, 182, 205-206
Bible, 113, 123, 174, 183
Black, 72, 125
Blackout, 57-59, 61-65, 67-70, 145
Blue, 51, 75, 157-158, 161-162
Blue Galaxy, 157-158
Blue Men, 161
Boats, 93
Bodies, 15, 85, 90, 112, 141, 207, 210, 216

Body, 10, 16, 32, 38, 90, 92, 97, 100, 102, 113, 116, 118, 168, 171, 181, 200-201, 208, 210, 213, 216
Bolivia, 58
Bomb, 12, 34-35, 73, 130-131, 140-141, 150, 177
Bone, 7, 90, 169-170
Bones, 32, 92, 112
Book, 39, 45, 65, 80, 83, 86, 89, 91, 99, 101, 124, 174, 206
Books, 26, 65
Born, 90, 108-109, 154, 194, 200
Boston, 57
Bottled, 42, 44, 172-174, 183, 204
Brain, 41, 127, 148, 171, 175, 218
Brains, 40-41, 204, 218
Brass, 39, 174
Brazil, 95-96, 122
Bread, 37, 115, 149, 213
British, 87, 137
Brother, 74, 86, 94-95, 151-153, 190
Brotherhood, 109
Brothers, 69, 74, 93-95, 109, 133-136
Buddha, 164, 206
Bullets, 163
Bulls, 182

Buried, 166

Burn, 37, 46, 61, 81-82, 84, 130, 153, 217

Business, 7, 15, 17

Butterfly, 67

Cable, 61-62

Calculations, 62, 104, 167, 170, 181

Calendar, 155

California, 22

Canada, 57, 145

Cancer, 127

Cancun, 183

Cannibals, 121-122, 128, 146

Cape Finisterre, 5-6

Capital, 38, 71, 94, 114, 173, 192, 204

Capitals, 162

Captain, 8-10, 15-16, 122-123, 144, 190

Capture, 74, 121, 128, 163, 165, 168, 207

Car, 37, 58, 99, 101-103, 105-107, 156

Caracas, 22

Carbon-14, 2

Caspian Sea, 198

Cataclysm, 66, 94-95, 157, 200

Cataclysms, 154

Catastrophe, 22, 30, 66, 92, 95, 131, 151-152, 158, 161, 168

Cause, 13-14, 29, 35, 61, 73, 87, 130, 147, 169, 174, 197

Center, 75, 81, 107, 119, 150, 152, 200

Centers, 76, 90, 171-172, 182, 201, 204, 218

Cerebrums, 60-61

Chain, 25, 209

Chaos, 29-30, 59, 168

Chile, 22, 58, 127

China, 127

Chosica District, 71

Christ, 16, 38, 63, 67, 143, 191, 201, 206-207, 218

Christian, 111, 133

Chronometers, 62

Chrysalis, 65, 67

Church, 86

Cibeles, 43

Cinema, 72, 182

Circumference, 75

Circumstances, 21-22

Circus, 79-80

Cities, 7, 12, 22, 25, 29, 51-52, 57-58, 63, 70, 87, 94-95, 97, 99, 104, 107, 110, 124-125, 129, 140, 146, 157-158, 160, 162-163, 173, 182, 192, 199, 203

Citizen, 34, 62, 95, 106-107, 109, 115

Civilization, 1, 21-22, 25, 30, 50, 60, 66, 74, 86, 106, 108, 137, 148, 156, 160-161, 198-199

Clairvoyant, 81, 214

Clarion, 159

Climate, 29, 72, 78, 104-105

Coal, 28

Coccyx, 169

Cold, 26, 82, 135

Cold War, 130, 193-194

Colombia, 22, 136

Colonize, 132

Color, 10, 51, 71-72, 75, 104, 125, 136, 211

Columbus, 5, 191

Comet Kondoor, 167, 169, 171, 173, 175, 177, 179, 181, 183, 185

Commission, 110, 130, 168, 170

Compassion, 66, 69, 206

Comprehend, 9, 16, 18, 21, 44, 49, 64, 67, 69-70, 117-118, 134, 137, 144, 167, 179-180, 194, 204, 214

Computers, 60-61

Concentrate, 91, 152, 214

Concept, 11, 13, 55

Conceptions, 6, 185

Concepts, 4, 10, 25, 34, 45, 49-50, 142

Conceptual, 201, 205

Congress, 130, 136

Connecticut, 58

Conquer, 3, 5-6, 12, 17-18, 20-21, 37, 129, 131, 133-134, 137, 140, 187, 189, 195, 207, 209

Conquistadors, 199

Conscious, 6, 116-117, 143, 202, 205

Consciousness, 31, 42, 44-45, 47, 49-50, 69, 115-116, 143-144, 147, 172, 174, 197, 200-202, 204, 208, 214

Contact, 23, 54, 56, 74, 110, 135, 143, 152, 160, 163, 189-190, 193

Continent, 167, 212

Continents, 30, 153

Contradictions, 39, 41, 119, 175, 203

Copper, 10

Corpse, 34, 165

Cortes, 140

Cosmic Mother, 179

Cosmos, 21, 49, 53, 55, 76, 93, 97, 128, 134,

139-140, 148, 170, 189, 195

Count Saint Germaine, 49

Covetousness, 37

Crash, 84

Craters, 23, 56

Create, 19-20, 30, 68, 90, 105, 199-200, 207, 211

Creation, 63-64, 67, 211

Creative, 28, 117

Crew, 8, 10-14, 16, 52, 71-72, 85, 96-97, 121-122, 165, 192

Crime, 17, 34, 39, 174, 218

Cross, 59, 154, 164, 183

Crucified, 164, 218

Cybeles, 180

Cyclops, 180

Cylinders, 171

Damage, 29, 61, 127, 130, 149, 156

Dangerous, 59, 61, 65-66, 84, 95, 161

Dante, 16, 174

Dead, 30, 56, 59, 118, 127, 192

Death, 47, 50, 118, 170, 184, 193, 209, 213-214, 216-217

Defect, 42-44, 117-118, 178-180, 204, 214

Defects, 12, 38, 42, 117-118, 135, 143-144, 172, 178, 204-205

Degenerated, 66, 133, 153, 199, 212

Degree, 93, 202

Delegations, 136

Deluge, 154

Demiurge, 28, 202-203, 206-207

Demons, 38, 172, 174, 202

Desire, 37, 113-118, 146, 200

Destroy, 30, 50, 69, 73-74, 96, 129, 131, 140-141, 145-146, 153, 157, 159-160, 163, 170, 199-200, 207

Destructive, 86, 96, 133, 139, 145, 147, 150, 153, 157, 161, 193, 199

Devi Kundalini Shakti, 44, 179

Devil, 14, 113, 118-120, 173

Devolution, 26, 28, 184-185, 200, 203, 209

Diana, 43, 180

Dimension, 18-20, 148, 200

Diogenes, 63, 67, 203

Disintegrate, 16, 42-45, 64, 118, 180-181, 204-205, 217

Disorder, 26, 29, 59, 110

Dissolution, 31, 114-115, 119, 135, 193, 216

Distance, 17, 28-29, 56, 102, 156

Divine, 8, 31, 43-44, 48, 134, 164, 179, 189, 201, 203, 206-207, 210, 216

Divine Architect, 48

Divine Mother, 43-44, 179, 210

Divinity, 46, 67, 75-76, 195, 203

Dogma, 19, 46, 207, 218

Dogue, 190-191

Doubt, 24-25, 29, 41, 57, 63, 95, 100, 104-105, 110, 138, 163, 173, 177

Dynamite, 54-55

Earthling, 15, 131, 159

Earthquakes, 22-23, 30, 36, 50, 125, 127, 153, 155, 168, 212

Eba, 190-191

Ecce Homo, 63

Echeverria, 87

Eclipse, 164

Economic, 188

Ecuador, 147, 156-157

Edison, 127

Ego, 44, 114, 175, 177, 185, 203-204, 216

Egotism, 35-36, 39

Egypt, 172, 199, 207

El Desierto, 7, 20, 143

El Salvador, 127

Elder, 8, 11, 14, 93-95

Electric, 59-62, 210

Electrical, 57-58, 63-64, 68-70, 103, 145

Electricity, 58, 60-62, 70, 180, 210-211

Electromagnetic, 130

Electronic, 60-62

Element, 149, 173

Elements, 31, 49-50, 73, 106, 130, 172, 181-182, 203, 215

Eliminate, 69, 143-144, 177-182, 201, 214

Elohim, 203, 207

Embryos, 21

Emotional, 40-41, 171-172, 200-201, 218

Emotions, 119, 175, 217

Enemies, 120, 129, 194, 213

Energetic

Energy, 20, 25, 27, 57, 68, 79, 103, 106, 113, 118, 145, 152, 160, 168, 170, 188-189, 210-211

England, 110, 137

English, 87, 199, 201, 213, 217

Entropy, 26-27

Envy, 37, 39, 127-128, 143, 172

Epistle, 31, 130, 213

Epoch, 77, 79, 200, 212

Equator, 23, 28-29, 150, 153

Equilibrium, 91, 168, 177

Esoteric, 16, 199, 201-202, 204-205, 210, 214-216

Esotericism, 183, 202, 206

Essence, 118, 172-175, 183-184, 204-205, 208-209, 216

Eternity, 19

Ether, 81-84

Euclid, 18-19, 41

Europe, 5, 188, 192

Eve, 95, 157

Evidence, 89, 149, 175

Evil, 36, 114, 144, 172-174, 204, 217-218

Evolve, 184, 200, 209, 217

Expedition, 189, 191-192

Experience, 24, 100, 140, 144, 160, 205, 209

Experiment, 85, 131

Experiments, 27, 93, 150, 159-160, 193

Explore, 36, 50, 118

Explorer, 121-122

Explosion, 24, 131

Explosions, 95, 118, 124, 126-127, 194

Expulsion, 81-82, 84-85

External, 45-46, 82, 142

Extraterrestrial, 5-6, 12, 19, 63, 66, 68-69, 72-74, 80-81, 139, 159

Extraterrestrials, 16-17, 20-21, 50, 137, 139, 141, 143-147, 149, 151, 153, 155, 157, 159, 161-166

Eyeglasses, 88

Eyes, 104, 160, 172, 176

Eyewitness, 52, 72

Ezekiel, 207

Face, 11, 30, 53, 55, 65, 123, 125, 154, 158, 182, 191

Fact, 7, 33, 53, 58, 66, 77, 115, 119, 143, 170, 178, 185, 209, 212, 218

Factories, 107, 148

Factors, 34-36, 193

Factory, 42, 71, 178

Facts, 12, 38, 49, 54, 121, 127, 141, 162, 165, 184-185

Factual, 185

Faculties, 108, 154, 177
Failed, 16, 74, 78, 94, 96, 133
Failure, 38, 185, 187
Fall, 24, 91, 125, 167
Falling, 84, 90-91
Falls, 90, 175, 218
Families, 108, 115
Family, 99-100, 108, 115-116
Famous, 34, 79, 122
Fantasies, 72, 97
Fashion, 77
Fate, 140, 162-163, 165
Fear, 35, 113, 146, 197, 209, 213, 217
Fifth Sun, 30, 155
Fire, 23-24, 30-31, 33, 43, 46, 153, 155, 158, 169, 200-201, 206, 210, 212, 217
First World War, 177
Fish, 27, 107
Flags, 109
Flame, 46, 113, 205
Flaming, 37, 43, 180, 205
Flammarion, 138
Flying Spheres, 81, 83, 85-87
Forest, 8, 11, 111
Forge, 133, 180-181, 205
Formula, 48
Formulation, 90-91

Fornicate, 18
France, 51-52, 127
French, 109
Friend, 15, 37, 69, 151, 155-156, 166
Friendly, 74, 76, 161
Friends, 57, 73-74, 117, 137, 155
Friendship, 74, 76, 190
Fruits, 27, 107
Fuel, 20, 59, 81, 84, 103, 188
Future, 3, 20, 33, 87, 181
Galaxy, 10, 14, 79-80, 141, 144, 157-158
Galilee, 164
Galileo, 127
Ganymede, 148-152, 154
Garden, 147, 152, 156
Gas, 58, 82
Gasoline, 58, 84
Gelatinous, 141
General, 10, 60, 207
Generals, 145
Genitalia, 171
Geometry, 18-20
Germ, 9
Germaine, St 49
German, 110, 201, 217
Gluttony, 38, 118-119, 143, 172, 194
Gnosis, 119-120, 157, 205-206

Gnostic, 54, 65, 74, 86, 111, 129, 131, 133-136, 155, 181, 195, 198, 202-203, 207, 212, 214, 218

Gnostic Anthropology, 155, 203, 207, 212

Gnostic Congress, 136

Gnostic Movement, 54, 129, 131, 133-136, 195

Gnostics, 46, 111, 135

God, 31, 45, 67, 173, 190, 198, 200, 202-203, 205-206, 210, 212-213, 216-217

God-Mother, 43-44

Gods, 13, 15, 197, 206-207

Gold, 11, 190, 208

Golden, 30, 46, 65, 75

Golden Blossom, 65

Gospel, 38-39, 63, 173-174, 200

Government, 60, 63, 68, 107-109, 115, 140, 163, 194, 212

Grafting, 27

Greece, 199, 210

Greed, 115, 118-119, 143, 172, 194

Greek, 124, 180, 198, 201-203, 205, 210, 216

Green, 51, 71-73, 75

Gringos, 101, 129

Guardian, 71

Guatemala, 22, 30

Guidance, 188, 198

Guinea, 12, 23, 121

Guns, 163

Hallucinate, 53

Happiness, 109, 111-112, 114-115, 119197

Hariton, 92-93

Harmony, 167, 174, 176

Hate, 39, 165, 176

Hatred, 73, 176, 195

Heart, 38, 70, 171, 187, 189, 192, 214

Heartbroken, 88

Heat, 31, 82, 125, 130

Heaven, 200, 213

Heavenly, 85

Heavens, 31, 130, 150, 152

Hebrew, 124, 205

Helicopters, 84

Help, 66-67, 74, 93-94, 96-97, 145, 152, 157, 162

Helped, 33, 86, 94, 96, 110

Helping, 103, 157, 162

Hercolubus, 153

Hermes, 16, 206

Hermes Trismegistus, 16

Himalayas, 95, 135, 198

Hiroshima, 177
Hitler, 73
Holocaust, 25
Holy Gnostic Church,
    86
Home, 47, 108, 122, 155
Homes, 108, 136
Homosexuality, 153
Homunculi, 65
Honduras, 22, 29
Honor, 99, 166, 198
Hope, 33, 152
Human, 11, 15, 21, 43,
    53, 63-67, 70, 72,
    77, 88, 97, 106, 111,
    120-121, 127, 135,
    142, 153, 155, 161,
    167-172, 201, 203-
    204, 206-209, 215,
    217-218
Human Being, 21,
    63-65, 67, 70, 106,
    120, 142, 153, 170,
    172, 201, 206-209,
    217-218
Human Beings, 53,
    63-67, 70, 77, 88, 97,
    120, 127, 155, 161,
    168, 204, 207-208
Human-Angels, 92,
    111-112
Humanities, 13, 94-96,
    131, 133-134, 163
Humanity, 10, 15, 30-
    31, 33, 55, 72, 87, 89,

93-94, 109, 116-117,
    129-130, 138, 144,
    153-155, 158, 163,
    167-168, 174, 193,
    205, 207-208, 218
Humanoid, 42, 67, 218
Humans, 11, 65, 97,
    112, 141, 157-158
Humility, 135-136
Hunger, 27, 35, 108,
    113, 127, 176
Hungry, 44, 106
Hurricanes, 29
Hydrogen, 73, 130-131
Hyperborean, 77
Ice, 23, 29, 82, 157
Icebergs, 23, 29
Idiosyncrasy, 33
Il Duce, 34
India, 198, 210
Individual, 10, 33, 38-
    39, 120, 167-168,
    170, 178, 181, 185,
    197, 203, 207, 210
Individuality, 119-120,
    173
Inferior, 148, 169-171,
    175, 215
Infernal, 184, 216
Infinite, 4-5, 9-10, 14,
    17-18, 20-21, 28, 50,
    53, 68-69, 93-94,
    122, 128, 131, 133-
    135, 137-138, 140-

141, 157, 189-190, 202, 204
Inhabit, 153
Inhabitable, 104-105
Inhabitant, 161
Inhabited, 4-6, 10, 13-14, 18, 109, 138, 140, 212
Initiate, 43, 94, 184, 194
Initiation, 16, 200
Inner Mind, 46-50, 142-143
Insoberta, 43, 180
Instinctual, 40, 129, 171, 201, 218
Insurgentes, 110
Intellect, 117-118, 185, 208-209
Intellectual, 6, 10, 40-42, 63, 65-66, 69, 119-120, 123, 167, 171, 201, 205, 207-209, 218
Intellectual Animal, 10, 63, 65-66, 69, 120, 167, 207-209
Intelligence, 6, 42, 50, 124, 162, 209, 218
Intergalactic, 10, 144
Intermediate Mind, 46, 142
Internal, 86, 116, 216
International, 111, 136

Interplanetary, 21, 79, 89, 111, 135, 188, 190, 193
Intuitive, 9, 197
Invade, 35, 66, 140-141, 194
Invent, 86-87, 140-141
Inventor, 73, 130
Invisible, 172, 203
Iraq, 127
Iron, 39, 58, 174, 207
Isis, 180
Island, 58, 154, 212
Italian, 34
Italy, 199
J. A. B., 111
Jalisco, 94
Japan, 127
Jealousy, 175
Jersey, 58
Jesus, 16, 38, 67, 111, 143, 173, 200, 207-208, 213
Jet, 74, 81-82, 84-85, 145
Jewel, 150
Job, 16
Jobs, 218
Journalists, 62
Judge, 42, 73
Jungle, 146, 188-190, 192
Jungles, 96, 160-161
Jupiter, 122, 148, 150, 156, 159, 201

Jupiterian, 121, 123-
    125, 127
Justify, 43
Kabir, 38, 173, 210, 213
Kali, 77, 79, 210
Kali Yuga, 77, 79
Kalkian, 77, 79
Kansas, 51
Kant, 45
Karma, 13-14, 114, 119,
    197, 215-216
Kill, 18, 35, 49, 128, 146,
    216
Kilometers, 7, 16, 55-57,
    59, 71, 101
Kind, 34, 100, 115, 206
King, 63-64, 67, 157,
    208, 213
King of Nature, 67, 208
Kingdom, 63, 67, 69,
    173, 184, 200, 209,
    213-214
Kings, 94, 110
Knowledge, 50, 70, 124,
    142, 193, 202, 205
Kondoor, 167, 169, 171,
    173, 175, 177, 179,
    181, 183, 185
Koosh, 198
Kraspedón, 122-124,
    127-128
Kundabuffer, 167-168,
    170-172, 181
Kundalini, 43-44, 179,
    210-211

Königsberg, 45
Laboratories, 110, 147,
    163
Laboratory, 12, 166, 189
Lamas, 95, 135
Languages, 96-97
Lantern, 63, 67
Lapides, 87
Laredo, 101
Laser, 141
Latin, 6, 187, 198-199,
    205, 215
Latin America, 187, 199
Law, 4, 26-27, 89-91,
    181, 197, 211-212,
    215
Law of Falling, 90-91
Law of Seven, 181, 211
Laws, 21, 134-135, 184
Layer, 124-125
Layers, 168-170
Laziness, 118, 143, 172
Legion, 38-39, 42, 113,
    118-120, 173, 204,
    209
Leibnitz, 9, 215
Lemuria, 212, 214
Lemurian, 77, 167, 212
Leones, 7, 20, 143
Lesbianism, 153
Leukemia, 127
Liberate, 45, 204
Liberated, 45, 204-205
Liberation, 182, 207
Lilliputians, 141

Lima, 71
Lingam-yoni, 180
Liquid, 20, 23, 79, 188
Loga, 190-191
Logic, 6, 80
Logical, 2, 56, 96, 117, 133, 187
London, 97, 162
Looisos, 170
Love, 11, 31, 38-39, 43, 53, 74, 108-109, 119, 122, 145, 176, 206
Luke, 173
Lunar, 1-2, 138, 169-170, 200, 216
Lungs, 39, 58, 174
Lust, 37, 116, 118, 143, 172, 179
Lustful, 39, 176
Machiavellian, 11, 140
Machine, 163, 168, 171-172, 175
Machines, 86, 96, 138, 189
Madness, 37, 131, 176
Magdalene, 38, 173
Magical, 180
Magnetic, 29, 92, 153, 210-211
Man, 5, 10, 31-32, 35, 63, 92-93, 99-102, 123, 147-148, 151, 155-157, 160, 173, 175, 200, 203, 207-209, 213

Manas, 183, 207, 213
Manasic, 173
Mantua, 30, 38, 174
Marah, 180
Marconi, Guillermo, 54, 160, 188
Mark, 108, 173
Marmande, 52
Martian, 76, 161, 188-190, 192
Martians, 54, 111-112, 138-140, 160, 188-193
Martinelli, 191
Marxists, 194
Mary, 38, 43, 173
Mary Magdalene, 38
Mass, 133, 153
Massachusetts, 58
Master, 174, 200, 216
Masters, 24, 48, 207
Mastery, 188
Materialistic, 142-143, 156, 165
Materialists, 194
Maternity, 108
Mathematics, 73, 124, 157
Matrimony, 65, 89, 214
Matter, 7, 17, 20, 34, 36-37, 53, 55, 59, 64, 68-70, 73, 110, 118, 135, 144-145, 166, 177, 185, 189, 197, 210

Matthew, 111
Mature, 93
Maturity, 108
Mayan, 183, 200, 213
Medicine, 1, 149, 170
Medieval, 6
Medina, Salvador
    Villanueva, 99-101,
    111, 155
Meditation, 111, 118,
    180, 204, 214
Melchisedeck, 67, 213
Memories, 113, 116
Memory, 48, 155
Mental, 108, 119, 172-
    174, 177, 202
Mentally, 99, 167, 177-
    178
Mephistopheles, 135
Mercurians, 111-112
Mercury, 56, 97, 111,
    149, 159
Mexican, 99, 101, 103,
    105, 107, 109, 111
Mexico, 7, 22, 51, 94-95,
    99, 101, 110, 140,
    164, 195
Mexico City, 7, 51, 99,
    110
Midwest, 51
Militant, 74
Military, 51, 60, 126,
    129
Minds, 73, 142, 175, 183
Mineral, 103, 184, 209

Minotaur, 44
Miscalculation, 170
Missiles, 129, 163
Mission, 2, 87, 137, 139,
    141, 143, 145, 147,
    149, 151, 153, 155,
    157, 192
Missionary, 111
Molecular, 110
Monad, 9, 201, 208,
    215-216
Money, 37, 106, 115-
    116, 119, 148-149,
    176, 187-188
Moon, 1-2, 4, 21, 28, 56,
    79, 122, 125, 129,
    133, 138, 149, 188,
    192
Mother, 9, 43-44, 107,
    109, 179, 210-211
Motor, 40-41, 171, 201,
    218
Mountain, 52, 102
Mountains, 65, 105,
    111, 146, 198, 203,
    207
Movement, 82-84, 107,
    113, 129, 131, 133,
    135-136, 195, 197
Movements, 54, 81, 87,
    111, 133-134
Multidimensional, 18
Multiple, 39, 141, 150,
    203, 207

Multiple Perfect Unity, 203, 207

Multiplication, 127

Multiplicity, 39, 41-42, 173, 203

Mussolini, 34

Mysteries, 47, 180, 197, 200, 210, 213

Mystical, 4, 118-119, 124, 134

Márquez, 87

Nagasaki, 177

NASA, 79-80, 132, 138-139

Nature, 21, 24, 27, 47, 49, 67, 77, 114, 148, 157, 181, 184, 197, 200, 207-209, 211, 215-216

Navigate, 73, 135, 137, 140-141, 144

Navigation, 90, 188, 190, 193

Navigator, 5, 125, 131

Nemesis, 181

Neptune, 159

Nevada, 54, 110

New Hampshire, 58

New Jersey, 58

New York, 57-59, 61, 63-65, 67-69, 97, 145-146, 162

News, 51-52, 54, 71, 87, 121, 127

Newspaper, 51, 71, 119, 127

Nicaragua, 22

North America, 54, 62, 110, 145, 156, 163, 199

North Pole, 23, 29

Nuclear, 1, 20, 125-126, 129, 193, 195

Núñez, Alberto San Roman, 71

Objective, 47, 67, 77, 80, 167, 170, 216

Observation, 34, 37, 71, 81, 171, 192

Observe, 11, 17, 36, 41, 52, 79, 175-176

Observed, 34, 52, 150, 217

Ocean, 5, 22-23, 33, 130, 149, 168, 212

Oceans, 22, 30, 33, 153

Oklahoma, 51

Old Testament, 183

Ontario, 57

Opinion, 4, 41, 85, 120

Orbit, 144

Orbital, 129

Orbiter, 78

Orbiting, 9

Organ, 167-168, 170-172, 181, 210

Organic, 171, 175, 181, 210

Organism, 169, 215-216

Organisms, 103, 169
Ors, 9
Ottawa, 57
Oxygen, 56
Pacific Ocean, 22, 33, 130, 212
Panspermia, 4, 216
Paraguay, 58
Paris, 97, 146, 162
Paropamisan Mountains, 198
Parsifal Unveiled, 65
Passions, 69, 113-114, 116
Pasteur, 127
Path, 16-17, 31, 101, 144, 158, 205
Patience, 55
Peace, 34-36, 111, 167, 186, 190, 193-194
Pennsylvania, 58
Perceive, 41-43
Percent, 76, 100, 106, 116-117, 204
Percentage, 204
Perception, 46, 117, 202
Perceptions, 45-47, 142
Persephone, 159
Persian, 113, 198
Peru, 71, 140
Perversity, 11, 31, 96, 133-134, 147, 218
Peter, 31, 130
Pharisees, 142-143

Philips Laboratories, 110
Phoenix, 139
Phosphorus, 127
Photograph, 52-53, 55-56, 138
Physical, 32, 41, 97, 108, 113, 116, 168, 172, 213, 216, 218
Physics, 1, 19, 25, 124
Pig, 12, 34, 121
Pilate, 63
Pilots, 29, 82, 85-86, 88, 102
Pineal, 148
Pituitary, 148
Pizarro, 140
Planes, 68-69
Planetary, 81-82, 91, 94, 131, 134, 153, 160, 163, 206, 213
Plant, 13, 60, 68, 209
Plants, 59, 61, 110
Plurality, 39, 41, 138
Pluralized, 39, 113-119
Pluto, 149, 159
Poet, 30, 38, 174
Polar, 77
Polarities, 70
Pole, 23, 29, 211
Poles, 28-29, 153
Police, 51, 194
Political, 109
Popluation, 198
Populace, 206

Populate, 65, 104

Populated, 18, 168

Populations, 58

Portuguese, 124

Power, 11, 43, 55, 60, 68, 103, 162, 170, 179-181, 202, 208, 210

Powered, 106

Powerful, 1, 24-25, 60, 70, 84, 145, 148, 156, 188, 198-199, 212, 214

Powers, 86, 180, 212

Practical, 42, 100, 117

Pride, 38, 118, 131, 133-134, 143, 172, 188

Priest, 189, 213

Prime Minister, 187

Principle, 43, 91, 208, 211

Principles, 31, 36, 38, 50, 211

Prisoners, 121

Problem, 61, 91-92, 116, 168

Propaganda, 34, 36, 195

Prophecy, 130

Prophets, 163

Prostitute, 35, 194

Pseudo-esotericist, 77

Pseudo-science, 79

Pseudo-scientist, 77

Psyche, 167, 181

Psychiatrists, 99

Psychic, 118, 149, 172-174, 177-178, 180-183, 204, 209, 214

Psychically, 167

Psychological, 22, 26, 33-34, 38-39, 41-45, 50, 143-144, 172, 178, 180, 201, 203-204, 214, 217-218

Psychologically, 33, 36, 151

Psychology, 32, 37, 39, 178, 217-219

Public, 54, 100, 144, 146, 194, 208

Publicly, 203

Puebla, 164

Quebec, 57

Race, 69, 116, 122, 133, 136, 154-155, 161-162, 197-198, 200, 212

Races, 72, 154, 198-199, 212

Radar, 51, 53, 59-60, 86

Radiation, 125, 127

Radio, 58, 72, 86

Radioactive, 25

Realities, 42, 45, 143

Reality, 1, 3, 6, 12, 14, 38, 41-42, 45, 47, 55-56, 62, 85, 89, 108, 138, 141-142, 147, 184, 210

Reason, 7, 33, 45, 67, 70, 122, 147, 182, 206

Reasoning, 6, 77, 80, 209

Red, 51, 72, 125, 172, 174, 218

Regenerate, 55

Region, 54, 110, 190, 217

Regions, 90, 198, 203, 216

Reincarnate, 115

Reincorporates, 114

Religion, 46, 76, 189, 195, 205

Religions, 46, 107, 136, 205, 207, 209, 212

Religious, 46, 107, 189, 206

Religiousness, 189, 194-195

Remind, 60, 142

Remnants, 160, 212

Republic, 94, 156

Return, 31, 122, 151, 156, 192, 204, 210

Returned, 14, 108, 123, 144, 151, 154, 157

Returning, 146

Returns, 76

Revelation, 125, 178, 217

Revolution, 26, 50, 92, 153-154, 205-206, 209

Revolutionary, 39, 92, 217-219

Revolutionary Psychology, 39, 217, 219

Rhea, 43, 180

Rhode Island, 58

Rockets, 1-2, 4, 20, 52, 55, 68, 79-80, 129, 133-134, 187-188, 193

Roman, 71

Romans, 31

Rome, 199

Roof, 71

Roofs, 75-76, 106

Root, 133, 154, 198, 200, 204, 212-213

Root Race, 133, 154, 198, 200, 212

Rotate, 46, 64, 150, 159

Rotates, 4, 82, 84, 150

Rotating, 29, 81, 83-85

Rotation, 81-82

Russia, 86, 127, 140, 193

Russians, 129

Sacrifice, 28, 30

Sacrificed, 210

Sadducees, 142-143

Sahara, 28

Saint, 49, 90-92, 183, 218

Saint Andrew, 183

Saint Germaine, 49

Saint Venoma, 90-92

Sainthood, 207

Saints, 37-38, 182, 207-208

Sakaki, 168, 170

Salvation, 134, 213

Samuel, 183

San Francisco, 22

Sanctuaries, 136

Sanctuarium, 111

Satan, 113, 201

Satellite, 122, 138, 148, 150

Satellites, 150

Satire, 76, 127

Saturn, 159

Saucer, 52, 71

Saucers, 4, 51, 53-55, 71-73, 75, 81, 86-88, 121, 129, 133, 135, 137

Saul, 183

Savagery, 121

Savages, 74, 85, 122, 212

Save, 95, 213

Saved, 33, 158

Saving, 107

Saviors, 145

Science, 49-50, 55, 70, 72-73, 77, 79-80, 85-86, 106, 133, 135, 142-143, 164-165, 190, 205, 211

Scientific, 2-3, 54, 68, 73, 89, 105, 124, 151, 160, 188-189, 191, 193

Scientifically, 70, 107

Scientist, 54, 110, 122-124, 160, 188

Scientists, 3, 25, 54, 79, 85-86, 110, 125, 138, 142, 149, 160-161, 188-192

Scriptures, 124

Second Death, 184, 209, 216-217

Second World War, 177, 199

Secrecy, 54-55

Secret, 84, 86-87, 95-96, 187, 205, 212, 215

Seed, 8, 198, 216

Seedgerms, 4-5, 216

Select, 56, 135-136, 154, 161

Self-criticism, 117

Self-discovery, 117, 178

Self-love, 53

Self-observation, 33, 41-43, 172, 178, 217

Self-realized, 64-65, 216

Self-realizing, 65, 216

Self-reflection, 178, 204

Self-willed, 42, 177

Sense, 5, 11, 18, 41-43, 50, 65, 92, 143, 148-150, 172, 206

Senses, 41

Sensory, 45-46, 142

Sensual, 45-47, 49, 142

Sensual Mind, 45-47, 49, 142

Seraphim, 90

Seraphim-angel, 170

Serious, 30, 91-92, 124, 182, 185-186, 218

Serpent, 180, 200, 210-211

Serpentine, 181, 210

Seth, 172, 174

Seven Spirits, 198

Sexes, 136

Sexual, 40, 171, 180, 201, 205, 217-218

Shakti, 43-44, 179

Sheep, 12, 163

Ship, 8-10, 12, 16-17, 52, 68, 72, 81-82, 84, 105, 122-123, 144-145, 147, 152, 156, 164-165

Ships, 1, 3-4, 6, 12, 18-20, 24, 53-55, 74, 79-81, 84-86, 88-93, 95, 97, 102, 107, 121, 134-135, 137, 140-141, 144-147, 149, 160, 162-163, 166, 188, 191-192

Simians, 169

Sin, 114, 216

Sins, 38, 114, 173, 204

Sir Weor, 86

Skeptical, 4, 76, 80, 127, 156, 187-188

Skeptics, 53, 100

Skin, 72, 102, 157, 208

Sleep, 102

Sleeping, 102, 210-211

Sleeps, 116, 204

Sleepy, 147

Slept, 203

Solar, 9-10, 20, 79, 103, 106, 125, 131, 133, 149-150, 152, 159-160, 188-189, 200, 207-208, 216

Solar System, 9, 131, 133, 149-150, 159

Somnambulists, 147

Soul, 113-114, 118-120, 205, 207, 209, 214-215

Souls, 112, 215

Sound, 197, 206

Sounds, 207

South America, 54, 136, 187, 190, 192

South Pole, 23

Southern, 96

Southwest, 52

Soviet, 130, 187, 194

Soviet Union, 130, 187, 194

Space, 1-6, 8-9, 14, 17-18, 20-21, 28, 36, 41-42, 45, 50, 52, 68, 73-74, 76, 79, 84,

90-95, 122, 125, 128-129, 131, 133-138, 140-141, 145-148, 153, 157, 187-190, 193, 195, 197
Spacecraft, 78
Spaceships, 52, 89-91, 93-97, 105, 110-112, 164, 188, 190
Spanish, 8, 101, 122, 199
Sphere, 81-84, 105
Spheres, 5, 81, 83, 85-87
Spine, 169-171
Spirit, 19, 113, 173, 190, 197, 200, 208, 215
Spirits, 173-174, 198, 206
Spiritual, 8, 94, 129, 136, 148, 198, 207, 210-211
Spiritualist, 100
Stability, 90-91, 168, 170
Stalin, 73
Steam, 23, 28
Steel, 8, 10, 81
Stella Maris, 43
Subconsciousness, 116, 118
Subjective, 77, 79-80
Submerged, 57, 184-185, 214
Suffer, 131, 175
Suffered, 57, 184, 213

Suffering, 31, 176, 197, 204, 216
Sufferings, 100, 184
Summum Supremum Sanctuarium, 111
Sun, 4, 30, 46, 64, 91-92, 107, 125, 155, 159, 212
Sunlight, 67
Suns, 90-91
Super-Man, 16, 50, 207, 213, 217-218
Super-civilized, 70, 76, 122
Superior, 43, 70, 113, 124-125, 130-131, 134, 171-172, 179, 200-201, 215
Superior Emotional, 172, 201
Superlative, 45, 47
Superman, 16, 50, 207, 213, 217-218
Supermen, 50, 141, 143-144
Survivors, 153-154, 165, 198, 200, 214
Symbol, 76
Symbolize, 75, 203
Symbolized, 203, 207, 210
Syracuse, 68
Tage, 161, 190-191
Tail, 82, 169-170, 201

Tanio, 140-141, 160, 192
Telepathy, 111
Telephone, 60
Telescope, 150
Television, 2, 58, 60, 72, 99, 152
Terrestrial, 11, 24, 94, 124-125, 129, 131, 138, 152, 167, 174, 191-192, 212
Tetradimensional, 19-20
Texas, 139
Theories, 4, 25, 49, 53, 142
Theory, 2, 26, 85, 205
Thermonuclear, 125
Thor, 19
Thought, 2, 14, 41, 45, 112, 146, 197, 217
Thoughts, 36, 214, 217
Three Brains, 40-41, 204, 218
Three dimensional, 18-19, 41
Tibet, 179, 198
Tibetan, 95, 200
Time, 5, 8, 18-21, 23-24, 26, 33, 42, 46, 49, 57, 69, 84, 87, 89, 92, 94, 116-117, 120, 127, 137-138, 145, 159, 170, 172, 183, 194, 212, 215

Titan, 150
Titans, 145
Tlaloc, 212
Tonantzin, 43, 180
Tower of Babel, 79, 131
Tranquility, 175
Transcended, 209
Transcendental, 47, 124, 180-181
Transform, 13, 50-51, 116, 125, 133, 175, 186, 215
Transformation, 28, 31, 50, 205
Transportation, 58, 85, 106, 109
Travel, 10, 14, 20, 50, 79-80, 93-94, 112, 134-135, 141, 188
Traveled, 99, 101, 103, 105, 107, 109, 111
Travelers, 10, 144, 157
Traveling, 82, 85, 153
Travelled, 160
Triangle, 105
Tribes, 122, 199
Tripod, 8, 10
Trojans, 4, 129
Truth, 12, 17, 45-46, 51, 53-55, 165-166, 186, 194
Tsunamis, 50, 153, 155, 168
Tube, 71, 81-82
Tuxedos, 86

Twins, 109
Tyranny, 206
Tyrians, 4, 129
Uan Weor, 86
UFO, 68
UFOs, 68
UN, 34, 36, 59
Unconscious, 54, 69, 147
UNESCO, 36
United States, 1, 51, 54, 63-64, 68, 86-87, 89, 101, 127, 130, 140, 145, 156, 163, 165, 187, 193, 199
Universe, 22, 24, 27-28, 45, 84, 90, 104, 142, 202-203, 206, 211, 216
University, 26, 48, 139
Unknowable, 19, 203
Uranus, 159
Uruguay, 58
Values, 38, 94, 205
Vanity, 143
Venoma, 90-92
Venusian, 106, 108
Venusians, 54, 103-105, 108-111, 115, 120
Vermont, 58
Vice, 12, 111, 175
Vices, 18, 69, 95, 114, 133, 153
Vigilant, 42, 117

Villanueva, 99-101, 111, 155
Violent, 115
Virgil, 4, 30, 38, 174
Virgin, 43
Volcanoes, 148, 153, 212
Vulcan, 159, 180-181, 205
War, 25, 36, 38, 42, 55, 108-109, 117, 121, 125, 130, 141, 177, 193-194, 198
Wars, 25, 30, 34-36, 95, 109, 193, 195, 199
Watchman, 42, 117
Weapons, 35, 68, 70, 86-87, 96, 133, 163, 207
Weor of Russia, 86
Whales, 183
Wheels, 81
White, 51, 102, 104, 122, 200
Wichita, 51
Wisdom, 11, 28, 48, 87, 108-109, 142, 144, 152, 205-206, 214
Woman, 119, 161, 175
Women, 35, 133-134, 173, 199
Word, 11, 13, 15, 18, 50, 65, 92, 143-144, 148, 150, 175, 190, 197, 201-203, 205-206, 210, 216

Words, 16, 34, 64, 102,
    137, 172, 174, 190,
    201, 205, 208, 214-
    215
Wore, 102
Work, 26, 86, 100, 115,
    148, 180-182, 184-
    186, 214, 217
Worked, 14, 152, 200
Worker, 202
Workers, 117
Working, 180
Works, 17, 34, 46-47,
    92-93, 106, 130, 188
World Wars, 25, 193,
    199
Worlds, 3, 12-14, 18, 66,
    116, 138, 159, 163,
    184
Writer, 10, 15, 26, 166
X-rays, 141
Year, 1, 54, 60, 89, 101,
    166, 187, 189, 191,
    195
Years, 20, 27, 71, 92,
    109, 137, 188, 200,
    209
Yellow, 72, 150-151, 159
Yellow Planet, 150-151,
    159
Yeshua, 173

To learn more about Gnosis, visit
gnosticteachings.org

Glorian Publishing is a non-profit publisher dedicated to spreading the sacred universal doctrine to suffering humanity. All of our works are made possible by the kindness and generosity of sponsors. If you would like to make a tax-deductible donation, you may send it to the address below, or visit our website for other alternatives. If you would like to sponsor the publication of a book, please contact us at 877-726-2359 or help@gnosticteachings.org.

Glorian Publishing
PO Box 110225 Brooklyn NY 11211 USA
Phone: 877-726-2359